Sustainability on Campus

Urban and Industrial Environments

Series editor: Robert Gottlieb, Henry R. Luce Professor of Urban and Environmental Policy, Occidental College

Maureen Smith, *The U.S. Paper Industry and Sustainable Production: An Argument for Restructuring*

Keith Pezzoli, *Human Settlements and Planning for Ecological Sustainability: The Case of Mexico City*

Sarah Hammond Creighton, *Greening the Ivory Tower: Improving the Environmental Track Record of Universities, Colleges, and Other Institutions*

Jan Mazurek, *Making Microchips: Policy, Globalization, and Economic Restructuring in the Semiconductor Industry*

William A. Shutkin, *The Land That Could Be: Environmentalism and Democracy in the Twenty-First Century*

Richard Hofrichter, ed., *Reclaiming the Environmental Debate: The Politics of Health in a Toxic Culture*

Robert Gottlieb, *Environmentalism Unbound: Exploring New Pathways for Change*

Kenneth Geiser, *Materials Matter: Toward a Sustainable Materials Policy*

Thomas D. Beamish, *Silent Spill: The Organization of an Industrial Crisis*

Matthew Gandy, *Concrete and Clay: Reworking Nature in New York City*

David Naguib Pellow, *Garbage Wars: The Struggle for Environmental Justice in Chicago*

Julian Agyeman, Robert D. Bullard, and Bob Evans, eds., *Just Sustainabilities: Development in an Unequal World*

Barbara L. Allen, *Uneasy Alchemy: Citizens and Experts in Louisiana's Chemical Corridor Disputes*

Dara O'Rourke, *Community-Driven Regulation: Balancing Development and the Environment in Vietnam*

Brian K. Obach, *Labor and the Environmental Movement: The Quest for Common Ground*

Peggy F. Barlett and Geoffrey W. Chase, eds., *Sustainability on Campus: Stories and Strategies for Change*

Sustainability on Campus

Stories and Strategies for Change

edited by Peggy F. Barlett and Geoffrey W. Chase

The MIT Press
Cambridge, Massachusetts
London, England

This book was set in Sabon by Binghamton Valley Composition in QuarkXPress.
Printed and bound in the United States of America.
Library of Congress Cataloging-in-Publication Data

Sustainability on campus : stories and strategies for change / edited by Peggy F. Barlett and Geoffrey W. Chase.
p. cm.—(Urban and industrial environments)
Includes bibliographical references and index.
ISBN 0-262-02560-4 (hc : alk. paper)—ISBN 0-262-52422-8 (pbk. : alk. paper)
1. Campus planning—Environmental aspects—United States. 2. Universities and colleges—Environmental aspects—United States. 3. Environmental management—United States. I. Barlett, Peggy F., 1947– II. Chase, Geoffrey W. III. Series.

LB3223.3.S89 2004
378.1'96'1—dc22 2003065130

Printed on recycled paper.

10 9 8 7 6 5 4 3 2 1

To John, Beth, and Sandy
and all the other students and teachers who will help shape our future

Contents

Acknowledgments

We are very grateful to the authors who shared their stories with us and endured our suggestions with such good will. Raquel Rojas, Lorraine Ulanicki, and Jennifer Smith helped us in the preparation of this manuscript, and Kathleen Derzipilski did a fine job with the index. We particularly appreciate the thoughtful guidance of Clay Morgan, our editor, and the three external reviewers; we believe the book is stronger for their wisdom. To Tony Cortese, Nancy Gabriel, Cynthia Staples, John Glyphis, Jeff Kuntz, Kim Shaknis, Tammy Lenski, and all the other folks at Second Nature, our sincere thanks for years of thoughtful encouragement of our own growth, as well as higher education's evolution toward sustainability. We also wish to thank Rick Clugston and Wynn Calder at University Leaders for a Sustainable Future for their work, their support, and for helping to secure funds that enabled us to get this project underway. The Atlanta HENSE meeting and especially Collette Hopkins's hard work deserve important credit for providing the inspiration for our efforts. We also want to thank the many friends and colleagues in addition to those we have already named, and those at our home institutions who supported our vision for this book by providing advice and support.

Collaborating on this volume has been challenging, intellectually stimulating, and joyful. Working from cities almost 3,000 miles apart, we have spent many hours on the telephone, exchanged countless e-mails, and found real delight in bringing this book to reality. May the many satisfactions of our own experiences in greening our campuses be shared widely, and may the projects in which you become involved be as sustaining as this book has been for us.

Introduction

Peggy F. Barlett and Geoffrey W. Chase

> Traditionally everyone [in the Pueblo culture], from the youngest child to the oldest person, was expected to listen and to be able to recall or tell a portion, if only a small detail, from a narrative account or story. Thus the remembering and retelling were a communal process. . . . Through the efforts of a great many people, the community was able to piece together valuable accounts and crucial information that might otherwise have died with an individual.
>
> —Silko (1996)

We have collected narratives from many colleges and universities around the country to preserve the "valuable accounts and crucial information" about the unfolding of a national movement toward campus sustainability. These stories inspire and delight—and instruct. They give us hope and remind us of our human failings; they bind us together in a discernable pattern within the chaos of everyday history. Most of all, these accounts of leadership illuminate the efforts of students, faculty, staff, funders, and administrators committed to a more sustainable future. In these chapters, we see the unfolding of institutional transformation as our nation begins to rethink how to live sustainably and in closer harmony with the natural world.

Our goal in collecting these stories is to provide ideas, options, and perspectives for those interested specifically in greening campuses, for those who want to reflect on how to change higher education, and for those who want to follow the unfolding of environmental awareness and cultural change in American institutions. The accounts here also provide strategies for those in higher education and elsewhere who experience a restlessness with the status quo and a desire to act. We trust these stories will serve as inspiration and nourishment for all readers.

Overview and Scope

The chapters and stories in *Sustainability on Campus: Stories and Strategies for Change* represent a remarkable diversity in higher education. The institutions themselves—four year, two year, public, private, exclusively undergraduate, doctoral granting, and research focused—reflect the range that characterizes higher education in the United States. The chapter authors are themselves a diverse group of faculty, administrators, a student, individuals with training in environmental studies, and others with a range of disciplinary backgrounds. Yet there are common themes and areas of focus that emerge from this diversity: environmental stewardship on campus, a focus on curriculum, an attention to green building design, engaging students and the community beyond campus boundaries, and system-wide initiatives. We have organized this book around those themes.

Readers will find particular chapters most relevant to their own situations and contexts, but our experience is that they can also draw useful lessons even from those stories that emerge from institutions unlike their own. Although not all institutions may be ready, for example, to contract for sustainable timber for construction of a new library several years away, such long-term thinking may spark a new set of ideas and strategies for bringing about positive change on other campuses. Even if a new building, new curriculum, or interdisciplinary research program seems unfeasible on one campus, the steps recounted here and the strategies used at other institutions may provide some wisdom for efforts that are indeed feasible or suggestions about how to move ahead with other initiatives.

In Part I, "Laying the Cornerstones," Christopher Uhl, a biologist at Pennsylvania State University, begins by describing the strategies for developing an environmental audit, which led to a much broader administrative stewardship structure at his institution. Abigail Jahiel and Given Harper, at Illinois Wesleyan, a small, four-year liberal arts college, recount early steps and missteps of the Green Task Force, a hard-working coalition of faculty who joined together with staff, students, and administrators to shift campus practices, especially with regard to recycling. Peggy Barlett at Emory University, a medium-sized, private, research university in Atlanta, discusses the importance of leadership and of collegiality in building a culture that recognizes the importance of environmental concerns in all aspects of the university.

Geoffrey Chase and Paul Rowland open Part II, "Redesigning the Curriculum," with a focus on the undergraduate curriculum at Northern Arizona University, and they describe how faculty revised more than 120 courses over a five-year period to include issues of environmental sustainability. Richard Norgaard, at the University of California–Berkeley recounts the history of the Energy and Resources Group, a pioneering interdisciplinary curriculum. His analysis examines how such an innovative effort at the graduate level can be maintained over several decades. Michigan State University, the prototypical land grant university with 33,000 students, is the setting for Laura DeLind's and Terry Link's work to shift campus awareness through an innovative course, "Our Place on Earth." Their experience of developing a single campus-wide course illustrates principles of sustainability and groundedness in place that echo sentiments in several other chapters. In the last chapter in this part, Debra Rowe, like Richard Norgaard, describes building political momentum for sustainability over an extended period. Her focus is on Oakland Community College, a large, two-year institution on the outskirts of Detroit with a very different mission from the University of California–Berkeley. She examines how the college moved over a ten-year period to adopt new curricular requirements ensuring that all students address issues of sustainability.

"Building Buildings, Building Learning Communities," Part III, consists of two accounts of efforts to bring about major changes in the way buildings on campuses are designed and constructed. In the first of these, David Orr describes the conception, planning, and construction of the Lewis Center at Oberlin College. His account of this pioneering green building reflects the collaboration of visionaries from across the nation as well as support (and some opposition) from within the college. His narrative, like that of Audrey Chang, a recent graduate of Stanford University, recounts the reshaping of institutional policies and priorities in design and construction of buildings. Audrey Chang describes this challenge from a student's perspective and discusses how a group of committed students can work with others in the campus community to make significant differences even in the policies of a well-established institution with a long tradition.

In Part IV, "Engaging Communities, Engaging Students," authors describe changes involving efforts that extend beyond campus boundaries. Richard Bowden and Eric Pallant describe how they and the students at Allegheny College engaged with the local business community to promote ecotourism in northwest Pennsylvania. Paul Faulstich also presents a picture of engagement with a broader community through his description of how Pitzer College students became involved in local grammar schools and the ensuing controversy over efforts to preserve undeveloped land. Also in California, but at a smaller two-year institution, Allen Franz of Marymount College, Palos Verdes, addresses stewardship issues by focusing on how students became involved in efforts to shift campus plantings away from nonnative and exotic plants to landscaping more appropriate to the geography of the region.

In the final part of *Sustainability on Campus,* "Building System-wide Commitment," four chapters address how the transformations we seek can help shape whole systems and the far-reaching impacts they can have. In addition, these chapters illustrate that shifts in the way institutions operate and function are not always top down or bottom up, but can be initiated from almost any position. Individuals in administrative positions, such as Patricia Jerman and her colleagues at the University of South Carolina and Polly Walker and Robert Lawrence at Johns Hopkins University, show how committed administrators can make key changes in how universities accept the responsibility of helping us move to a more sustainable world. Polly Walker and Robert Lawrence describe a university-wide initiative cutting across the schools at Johns Hopkins, and the South Carolina case explores how multicampus coalitions can be created and fostered. Michael Edelstein at Ramapo College in New Jersey, recounting a much longer history, shows that institutional gains happen unevenly, and sometimes almost serendipitously. His description of the Solar School House, the Alternative Energy Center, an ecological literacy program, and the emerging New Jersey Higher Education Partnership for Sustainability illustrates that real change is not easy but that it is nevertheless possible. Finally, Nan Jenks-Jay, at Middlebury College in Vermont, provides a glimpse of how to build systemic sustainability, broadening from academic engagement to major changes in campus dining and long-term purchasing policies. Her chapter shows what can be accomplished with collaboration and senior-level support after several decades of commitment to sustainability principles.

Together, these narratives show that higher education has the potential to be a critical leverage point for change. They also suggest that moving a campus forward means engaging in systemic shifts that inevitably bring stops, starts, successes, failures, and organizational learning. They further suggest that pursuing sustainability is not easier, or more difficult for that matter, at one type of campus as opposed to another. Perhaps most significant, the diversity of these accounts illustrates that the opportunity for leadership and the possibilities for change have more to do with how one sees oneself within an institution and in relation to others there than the position one holds.

In becoming aware of global environmental degradation, the growing gap between first and third world countries, as well as environmental challenges in our own communities, colleges and universities have an important role to play in shaping our future. Campuses across the United States alone represent an enormous investment in buildings and land, and therefore how we maintain and build our physical plant, engage in buying practices, dispose of waste, and consume energy is critically important to the environmental health of the broader society. Higher education's influence reaches beyond campus boundaries as well. Colleges and universities are inextricably woven into the communities in which they exist, and their programs, commitments, and connections provide opportunities to make significant differences off campus as well as on campus. Perhaps most important, and most obvious, colleges and universities in the United States teach approximately 14.5 million students each year, and these future citizens and leaders will play a critical role in helping us move to a more sustainable future.

We celebrate the courage of the authors of this book. They have stepped beyond the conventions of academic discourse to promote a new path, introduce a new idea, and midwife a more environmentally responsible future. They have been willing to share their successes and also the challenges they faced, the setbacks they experienced, and the understandings they gained. Thus, these chapters do not just recount successful stories about green campuses. They do not focus solely on *what* has been accomplished on different campuses. They focus on *how* those accomplishments were achieved. We believe these accounts of leadership, these strategies and tactics, these peaks and valleys, provide a rich history and guideposts for others who follow.

Sustainability

This book brings several decades of international engagement with concepts of sustainability and sustainable development into dialogue with the realities of higher education in the United States. As Thaddeus Trzyna notes in the Introduction to *A Sustainable World,* the term *sustainable development* originated in the 1970s but was popularized in the Brundtland Report, *Our Common Future,* which was published in 1987. Trzyna goes on to note that by 1995, only eight years later, at least seventy definitions of *sustainable development* were in circulation (1995, 23). The disagreement over the exact definition notwithstanding, we can point to central concepts and ideas that cut across most, if not all, definitions and these in particular capture how most of the authors included in this book define *sustainability.* The first and one of the most straightforward of these is one of the earliest: "Sustainable development is development that meets the needs of the present without compromising the ability of future generations to meet their own needs" (Brundtland 1987, 43). It is this definition that continues to be cited most often as people seek to understand this concept.

Additionally, most authors in this book understand sustainability to be the intersection of three domains: the economic, the environmental, and the social. The economic realm includes the production of goods and services to support the livelihoods of a population, while the environmental realm includes maintenance of biodiversity and the health of biological systems in a region. The social realm of the sustainability paradigm often includes social justice issues and in particular, broad political participation (Holmberg and Sandbrook 1992, 34). Seeking the intersections of these three realms requires complex trade-offs, and we note that even the World Bank in recent years has embraced a version of the sustainability perspective—"the triple bottom line." It is perhaps inevitable, that many of the authors in this book focus more strongly on environmental dimensions as a necessary first step toward sustainability, although attention to engagement, participation, and building community is also strong. Economic constraints and concerns emerge less often, but form a backdrop to the initiatives recounted.

Finally, from our perspective, sustainability is not a problem, not something to be solved, but rather a "vision of the future that provides us with a road map." This map "helps to focus our attention on a set of values and ethical and moral principles . . . to guide our actions" (Viederman 1995, 37). Viederman's approach echoes the three domains of sustainability, arguing that our societal goal is "to ensure to the degree possible that present and future generations can attain a high degree of economic security and achieve democracy while maintaining the integrity of the ecological systems upon which all life and production depend" (37). Sustainability in this instance is not an end point, not a resting place, but a *process*.

The process of sustainability begins with an awakening to emerging problems caused by conventional norms of behavior (both institutional and personal) and then a discernment of new directions without a specific sustainability checklist. As the chapters in this book illustrate, every institution of higher education has a particular history, mission, and pattern of action. Clarity about what constitutes sustainability for a particular institution will require the shared wisdom of many points of view. We find that the vision, an alternative sense of future, becomes reality through relationships—learning, questioning, trusting, competing, at times coercing, and at times building together. Through individual and collective action, these relationships bring about institutional change, though change does not come easily. The chapters in this book thus support the optimistic assessment made by Fischer and Hajer that the paradigm of sustainability can indeed foster institutional learning (1995, 3).

Michael Redclift highlights the journey toward sustainable development as having two parts. First, research on "current social processes" clarifies how we produce and consume and how power, meaning, and institutions—on many levels—affect the environment. With this understanding of our current situation, we then must "establish the feasibility of intervention in these processes. To facilitate this intervention, we need to know more about existing experiences of positive environmental action to establish the conditions of their success" (Redclift 1997, 265). This book is our contribution to Redclift's second charge, and it stands as a very early history of a movement in higher education that we predict will continue.

Sustainability within Higher Education

Environmental awareness on university campuses burst into public consciousness in 1970 through an Earth Day celebration during which students buried an automobile to symbolize the deleterious impact of humans on the environment. The subsequent energy crisis of the 1970s focused even greater awareness on environmental challenges; in addition to lowering thermostats and reducing gasoline use, national legislation emerged in the Clean Air, Clean Water, and Endangered Species acts. These shifts toward greater environmental awareness were echoed in various campus efforts to establish environmental studies departments, preserve green space, and promote hiking and outdoors activities as dimensions of physical education. Often led by students and staff, a few campuses pioneered more sustainable daily operations and long-term planning.

Internationally, awareness of ecosystemic threats was also growing. Amazonian deforestation, water shortages, acid rain, and the possibility of global climate change, among other issues, galvanized attention to sustainable development. Two years before the United Nations Earth Summit in Rio in 1992, a group of university leaders united behind the Talloires Declaration, adopted in Talloires, France <http://www.ulsf.org/programs_talloires.html>. This document recognized that "universities have a major role in the education, research, policy formation, and information exchange necessary" to address "environmental changes . . . caused by inequitable and unsustainable production and consumption patterns that aggravate poverty in many regions of the world" (1990). It was also understood that the unprecedented speed with which we are destroying and degrading our environment has dramatic implications for the quality of life for all people, now and in the future. Today, more than 300 institutions of higher education have expressed support for these sentiments.

In the years following the Talloires Declaration, a number of organizations and individuals worked to promote the role of higher education in sustainability efforts worldwide. One of these individuals, Tony Cortese, a principal author of the Talloires Declaration, recognized that though higher education is highly resistant to change, societal progress in sustainability requires a partnership with colleges and universities. Tony went on from Tufts University to found Second Nature, a nonprofit organization dedicated to promoting sustainability in higher education. Second Nature appears at critical junctures in several of the following chapters, and it provided the arena where issues—and stories—inspired both of us to begin sustainability efforts on our respective campuses. This book is strongly influenced by Second Nature, whose workshops emphasized the interplay of four campus domains of sustainability: teaching, research, outreach, and stewardship.

Other organizations, scholars, and activists have also played key roles in furthering sustainability efforts in the past fifteen years. University Leaders for a Sustainable Future (ULSF), under the direction of Rick Clugston, has been instrumental through its newsletter, *The Declaration,* and through workshops and symposia. Julian Keniry founded the Campus Ecology Program in 1989, through the National Wildlife Federation, to stimulate campus activities supporting the celebration of Earth Day 1990. Today, Campus Ecology works on more than 140 campuses to transform them into ecologically sustainable societies.

Yet another important support for the greening of higher education were the biannual conferences hosted by Ball State University in Muncie, Indiana. These formal gatherings provided the setting for faculty, groundskeepers, students, architects, and other campus actors to learn together about best practices and strategies at other institutions. On the institutional level, the *International Journal of Sustainability in Higher Education,* edited by Walter Leal Filho and published in Germany, gave American innovators contact with international collaborators and fostered even more awareness of options and innovations.

Sustainability on Campus: Stories and Strategies for Change has its roots in a broader history that we have only briefly recounted here, but the specific impetus for it emerged in Atlanta, Georgia, in 1999. Seeking a framework to jump-start innovation at colleges and universities around the country, forty people came together to explore a possible national network for sustainability efforts. We and several others looked particularly at how to foster campus leadership. We came to the mutual realization that stories of change and leadership at other schools could provide the invigorating fertile ground for the development of future campus action. We knew that our own campus efforts had relied on stories that explained key principles for a new paradigm of intellectual growth, environmental responsibility, social equity, and broad participation. Inviting narratives from among campus leaders we already knew, we also sent out an open call in an effort to seek stories from individuals we did not know. The responses affirmed our instinct that there are many important stories to be told—unfortunately, more than we could include in this book. At the same time, we were heartened to realize that there are so many stories in progress, and we look forward to future opportunities to broaden the conversation that we hope is strengthened by this collection.

Barriers to Change

Although higher education represents a powerful leverage point for societal change, its conservatism presents significant challenges. Among the most tradition-bound institutions in our society, colleges and universities provide security and familiarity in their ivy-walled image. The American Association for Higher Education (AAHE) documents that of the many initiatives introduced in the past forty years, few have taken hold, and most institutions, at least in terms of their basic curricular structure, operate today much as they did yesterday (Veysey 1965). It is worth discussing some of the barriers to change here to illuminate the challenges that the authors in this book have been working to overcome.

Disciplinary Boundaries

Disciplinary boundaries in higher education today are perhaps the single largest impediment to achieving any kind of substantive transformation or reorientation. Resources are linked to disciplines, and as many of our chapter authors lament, the structure of the institution gives rise to a status quo that militates against interdisciplinary work. Faculty members are rewarded primarily for the teaching and scholarship related to their disciplines, and they are discouraged from moving across disciplinary boundaries.

This is particularly problematic as we think about addressing issues of sustainability because this topic is by its very definition interdisciplinary in nature. Sustainability cannot be cordoned off into one area of the curriculum such as environmental sciences because it encompasses a range of disciplines, including the arts, humanities, economics, political science, history, other natural sciences, anthropology, as well as work in professional programs such as business, engineering, law, and health professions. Sustainability issues make for messy, complex research problems, requiring new professional skills and new criteria of evaluation. Thus, the reward structure of higher education, which is linked directly to departments, often discourages faculty from researching the broader issues involved in steps toward sustainability.

The disciplinary structures of our institutions foster courses that are usually content focused, leading students to understand learning as a means for acquiring knowledge instead of addressing issues, problems, or challenges. The very structures of our curricula, and the fact that they are "owned" by departments that exist as disciplinary strongholds, stand against the kind of pragmatic teaching espoused by Dewey and Meiklejohn (Dewey 1916, Meiklejohn 1981). It is precisely this kind of pragmatic approach to teaching that may be the most critical as we prepare students to face the challenge of creating a more sustainable future. Further, disciplinary and departmental divisions, characteristic of the majority of higher education institutions, work against interdisciplinary teaching and telic programs, two features identified by Grant and Riesman (1978) as central to an alternative paradigm for teaching.

Graduate training, designed to ensure qualifications for good jobs, can reproduce the status quo and solidify disciplinary boundaries. Even if students receive fairly broad training as undergraduates, once they enter graduate programs, they usually engage in training that encourages them to become increasingly focused and narrow. Richard Norgaard's chapter highlights an unusually strong example of interdisciplinary collaboration at the University of California at Berkeley, that such barriers can be overcome.

Silos and Scale

Our recognition of disciplinary boundaries leads us to acknowledge two related barriers. Using a metaphor borrowed from systems thinking, we note that academic culture tends to be organized into silos—insulated, vertical units with little cross-flow of information. Universities suffer from "numerous sub-cultures of decision-making styles, time constraints, priorities, and experiences" (Sharp 2002, 132), making communication across these silos a difficult challenge. Graham and Diamond point out that medical schools, in particular, are always separate "dukedoms" (1997). The National Wildlife Federation's survey of colleges and universities found "almost a complete disconnect" among various departments on any given campus (National Wildlife Federation 2001). Surprisingly, our chapters show that this problem affects not only the large universities but even the small liberal arts colleges. Institutional growth also leads in some cases to extreme burdens on faculty and administrators, and hence greater inefficiency in dealing with environmental problems or responding to sustainability opportunities (Sharp 2002, 133). The sheer scale of higher education makes change difficult, and yet this is also, as we noted earlier, what makes it so important.

Inspired leaders who choose to work across boundaries and to make connections among disciplines are not always successful. The role of the scholar-activist "may be time-consuming, exhausting, and discouraging, when the results of one's efforts are neither . . . apparent nor personally rewarding" (Baer 2002, 48). Further, many individuals lack key skills, and training in institutional change is rare. Lack of support from the institution or from peers presents the risk of being marginalized and discredited by the normal prestige markers of the school. Sustainability efforts require faculty and administrative leaders to decenter their own expertise and to accept "newcomer" status in what can be seen as an overwhelming intellectual task. Demands for rigor can silence creative investigation, as David Orr has pointed out (1993). Leaders must also learn multiple languages in order to interact effectively with diverse constituencies, as Rich Bowden and Eric Pallant and several other of our authors have described.

The conservatism of these institutions is further challenged by the complexity of the innovations required by sustainability. Our received wisdom about the quality of life, our attributions of value, even our ethical compass, are destabilized by the environmental and social challenges we now face. Because the steps we will have to take will require fundamental shifts in the way we think about our place in the world, we will increasingly think as much about what we propose to change as well as how we arrived at these changes. What is at stake is not only which technological innovations we choose but how we will choose them. But as this book demonstrates, these shifts are already underway, and many in higher education have begun to lay the groundwork for sustainability. When even much simpler educational innovations, such as the addition of driver education to high schools, took over eighteen years to become widespread (Rogers 1971, 64), we are called to extraordinary dedication to guide and support such profound change. Several of our chapters look at how we sustain ourselves in this process.

Financial Pressures and Public Responsibility

Disciplinary boundaries are linked to financial pressures within higher education, which can form a barrier of their own to sustainability efforts. Financial realities have forced some disciplines to be grants driven, and many universities have built liaisons with corporations and government agencies to support research as well. These complex linkages can foster social conservatism and weaken the legitimacy of societal criticism and social engagement.

Thus, while many colleges and universities recognize the role they play in preparing students for citizenship and in acquiring the skills and abilities that will enable them to work with their communities to address the larger problems facing society, this is not their primary focus. Bruce Kimball has argued in his history of the idea of liberal education that higher education in the United States since the late nineteenth century has become increasingly focused on research, on developing new knowledge and technology, and this is where the rewards are the highest (Kimball 1986). Indeed, courses and activities to apply academic knowledge to societal problems are often seen as less prestigious or even inappropriate, depending on the field. An atmosphere of "studied silence and subtle disapproval" can discourage younger scholars, and few professors "receive guidance on how to manage the public role of an academic" (Sabin 2002, B24). The tendency of many in academic institutions to shy away from public engagement "stifles badly needed dialogue on problems facing society as a whole" (Sabin 2002). In addition, education to engage the public in issues of sustainability may require a broadening of the definitions of scholarship, with attendant *tsuris* over definitions of excellence, implications for tenure, and expectations of the professoriate

(Boyer 1990). Aronowitz reminds us that the regimentation of professional careers keeps our noses to the grindstone and prevents us from making history (2001, 4).

In undergraduate institutions, time-consuming efforts to train students to carry out on-campus research or community projects may be supported by an ethos of commitment to mentoring and hands-on teaching. Faculty may not, however, feel rewarded for such extra effort, and often the wear and tear on faculty—and staff—can be considerable. In a research university, such teaching efforts may be supported by the presence of graduate teaching assistants, or they may be discredited as insufficiently prestigious or connected to major scholarly work. Some research institutions can carry out more sophisticated sustainability projects over a longer term than is possible in a liberal arts college, while it is more often the case that faculty judge that their careers will be harmed by the risk of such efforts.

Multiple Stakeholders

The many parties interested in higher education make change difficult as well. Unlike corporations, for which profit and the financial bottom line provide a firm metric of success, goals for higher education vary from stakeholder to stakeholder. Students, their parents, alumni, boards of trustees, local communities, and state legislatures all have valid claims on the energies of the school, and each must be a part of the transitions required for sustainability. Even where a president and university senate might be ready to adopt the Kyoto Accords guidelines for energy reduction, board of trustee members who are unconvinced may block the effort. The new learning of sustainability—and its disquieting loss of certainties—must be shared by all university stakeholders.

At the same time, political and economic pressures by donors, state legislatures, corporate interests, and local elites can work in favor of change toward sustainability. Chapter 13, on South Carolina universities, illustrates an initiative toward sustainability that emerged from an external funder. Other institutions have been pressured by the Environmental Protection Agency to clean up practices regarding hazardous chemicals or that contribute to air and water pollution. The Pitzer College chapter by Paul Faulstich describes an unusually successful partnership with community schools—as well as some political turbulence with aggrieved university stakeholders.

Place, Context, and Commitment

Our stories are drawn from just a few of the hundreds of colleges and universities in the United States whose experiences echo similar changes in Canada, Mexico, Europe, and other universities around the world (Leal Filho 2002). The institutions represented in this book cover the full range of the Carnegie Foundation classification system and include everything from community colleges to research universities. Some have endowments in the billions and others virtually no endowment at all. Some are highly selective in their admissions process, and others admit students with all levels of preparation. Some grant many degrees at the doctoral level and others only the associate of arts degree; the rest fill out the spectrum in between. These different kinds of institutions can be expected to embrace sustainability in uneven and contradictory ways (Roseberry 1994, 86). Social change is not a unilineal process, and we have chosen our stories to evoke such discontinuities as well as some parallels.

The place in which each of these institutions stands—the immediate environment—is also a key difference among them, and it plays a key role in our narratives. Sustainability, many would argue, requires a detailed knowledge of local areas and local actors, and environmental literacy requires knowledge of the campus in the context of its local ecosystem (Clugston and Calder 1999). Forms of engagement with the natural world will depend on the location and type of each institution. Faculty who choose a job at Northern Arizona University, for instance, cannot help but understand that the surrounding natural beauty of the nearby Grand Canyon is part of the context of their decision. Faculty coming to the suburban campus of Emory University often know about the cultural context of Atlanta and the South but are unaware, even after many years, of the ecological systems that surround and affect the campus. Opportunities to learn about the health effects of urban sprawl or to participate in the restoration of an urban forest are useful pedagogical exercises in the latter context in a way that may not be true in the former.

The challenge, then, of adopting systems thinking and awareness of place varies by locale.

Leadership: Sources of Inspiration, Personal Commitment, and Institutional Change

Given these barriers to change in higher education, it is somewhat curious that in so many diverse schools, teachers, scholars, researchers, students, and administrators have set out on this difficult path of systemic change. What was their source of inspiration? Some, like Chris Uhl, recount witnessing the devastation of the Amazon and acknowledging the connections between these events and our daily lives. A love of nature or bond with nature is revealed as a motivating force by several authors, including Allen Franz. Scientific knowledge of harms—to human health, as well as to ecosystems—galvanizes Polly Walker and Bob Lawrence at Johns Hopkins University. Some people become sustainability leaders because of a professional opportunity, a career shift that opens doors. Sometimes a convergence of forces—a new building, a new funding source, a damaging new road—galvanizes action.

Whatever the place and the commitment, we see that the changes brought about by these individuals take many forms. Two dimensions of these changes are of particular interest: new institutional routines and new socionatural relations (Fischer and Hajer 1995, 2). We see evidence in this book for both of these shifts as new curricular requirements for students, new course certification, or the adoption of construction guidelines all help establish new routines. In the Michigan State story by Laura DeLind and Terry Link, we see both students and faculty learning about the natural world around the university by deepening their connection to place.

Most authors have not articulated a radical critique of American society, but instead have focused on quiet ways to make alternatives palatable—as have Abigail Jahiel and Given Harper at Illinois Wesleyan—to prod the imagination—as David Orr has done at Oberlin—or to explore through research ways to construct an alternate future—as Nan Jenks-Jay has helped bring about at Middlebury College. Few authors fall into the trap of reducing sustainability to a scientific problem to be solved with risk management technology or improved engineering. Instead, they highlight the messy reality of participatory engagement in cultural transformation.

Some Lessons Learned

We recognize that "what really happened" at each institution described here is much more than we can compress into our stories. Our narratives are contestable accounts, glossing over conflicting interpretations of events. We know that human experience is always reordered and reinterpreted over time (Barth 1994), and though we do our best to capture relevant processes and experiences, our perceptions are always in flux. Sometimes, a "strategy" is more a post hoc deduction than a conscious choice at the time (Wilk 1989, 28). Persons in other positions at the same institution will write other accounts, and we welcome those voices to build a history of this momentous transformation. For the time being, however, we believe it may be useful to point to the most salient and powerful features that cut across most, if not all, of these stories.

Personal relationships are critical. Many of our authors document that sustainability efforts thrive better when personal contacts are used in place of more impersonal e-mails, newsletters, and announcements. Phone calls, lunches, personal warmth at a formal meeting—these dimensions of creating community have been shown to be more important than many of us realized at first. Though people come forward often because they care deeply about the environment, frugality, and ethics of sustainability, they stay and find meaning in the human friendliness and laughter, the deep satisfactions of knowing people "from across campus."

Trust, which emerges from strong relationships, drives the change we seek. All through the stories are the accounts of new relationships and connections across campuses. The linking of arms to create systemic changes called for by the vision of sustainability requires the slow building of trust. Faculty must join with staff and administrators and the other stakeholders for the long haul, bringing successive generations of students into the fold, creating self-renewing communities of learning. Although we emphasize the importance of trust, networks, and collaboration, competition among schools also plays a role in our stories. For example, Trish Jerman, Bruce Coull, Alan Elzerman, and Michael Schmidt recount how the rivalry among the three South Carolina schools was used effectively to invigorate attention to environmental issues. Social groups have hierarchies and conflicts, and thus, although many authors have softened those edges of our narratives, several of these accounts reflect an undercurrent of wanting to be recognized for good work. In several other cases, there is also a sense of wanting to be in a leadership position among institutions or to be perceived as equal to institutions with superior reputations.

Success is not always related to the numbers of people involved. We've learned from our stories that much can happen with only a few active individuals. At Stanford, for example, Audrey Chang and a small group of students led the administration of the university to make significant changes in campus building guidelines. At Penn State, Chris Uhl worked initially with only a few students to start what developed into a major university-wide initiative. At many institutions, one or two here and a handful there is all it takes to get started. At other colleges and universities, broad participation works effectively, and this pattern reflects what many see as an essential component of sustainability.

Many researchers stress that information flow to all stakeholders is key, but our stories show that regardless of the number of people involved how very difficult it can be to communicate. As we noted earlier, the scale and silos that characterize higher education, as well as the fact that they are loosely coupled organizations, makes them challenging sites for drawing many into the same movement or initiative. Fortunately, as a number of these narratives illustrate, rather impressive movement can occur without all the stakeholders being on board.

Different paths are fine. We see much variability in strategies and starting points. At some schools, such as Allegheny College, faculty initiated a variety of community development projects that enabled the college to envision a broader regional sustainability partnership, educating students in the process. It is also clear from this group of authors that we have to work from our strengths. Some of us are people oriented and do well encouraging broad participation. Others are more technology driven or like to develop policies and campus governance structures.

Just as there are different paths, each of those paths begins from different origins. Many of the stories in this book begin with recycling. It is visible, noncontroversial, and inherently virtuous to reduce waste. It even has the potential to make money for the college or university. But we can see from these accounts that recycling, though logical, can be a difficult way to start. It requires space, new staff or more work for old staff, and some significant costs for storage and transportation. Abigail Jahiel and Given Harper show that the habits targeted are complicated, and the population to be educated is constantly changing. It is a behavior change that requires long-term effort to sustain, and we conclude it may not be the ideal way to begin campus greening everywhere, though it seems to have worked at Illinois Wesleyan and at other institutions.

Others begin with measurement. They use campus audits to reveal many important opportunities for redressing environmental harms—"the low-hanging fruit." Geller (1994) reminds us that audits are a good measure of our priorities and institutional concerns and can be very effective forces for change. Other authors in this book focus on deep institutional values, such as strong relationships between faculty and students, that cannot so easily be measured or quantified. We see that Allen Franz's commitment to a delightful landscape affects many levels of decisions, without the collection of campus-wide data. Some offices on university campuses, such as purchasing, are easier to measure, as they seek, for example, to buy more environmentally friendly products. Areas with a command-and-control subculture, such as university operations or hospital administration, may be more amenable to clear mandates toward a sustainability revolution. Still others will want to focus on the intellectual or political momentum of curriculum. We need more research to assess the ways different institutions respond to incentives and constraints, and how best to foster more subtle behavioral changes in our institutions.

Leadership emerges from many different sources. Some of the chapter authors clearly are unusual, charismatic individuals. But as one author acknowledged about one of the founders of his campus effort, "He was not particularly charismatic, just determined and a good listener." In another case, a faculty leader had charismatic qualities in the classroom, but for other faculty his leadership came more from creating an intellectually dynamic program. In still other cases, significant change is catalyzed through the efforts of a student such as Audrey Chang or a staff member such as Trish Jerman. In some of these cases, individual faculty members took on key roles, while in other cases, at Northern Arizona and Allegheny, for example, small groups took on key roles. Although none of the chapter authors is an administrator of campus operations, such as facilities management or purchasing, these individuals and their central leadership roles are visible in many of the chapters.

Distinct personalities and positions can all make a contribution. Some faculty work at the center of campus (or even consortium) politics. Others work more at the edges of campus power, with courses, student projects, or community-college liaisons. Some, such as Terry Link, have a formal position as director of the Office of Campus Sustainability that enhances and legitimizes their advocacy, but many are volunteers, building coalitions out of a personal sense of mission and commitment.

Support from above is critical. Several of us point out that affirmation from the top is essential. The catalyst for change among some of the institutions represented here came from state governments, as in the case of New Jersey and South Carolina. At others, critical support comes from a president or another administrator. Polly Walker and Bob Lawrence recount the steps leading up to an eloquent public letter from the president of Johns Hopkins. But it is also clear from our histories that such support can be slow to come or be backed up by actions or budget, yet significant momentum toward campus greening can nevertheless take place. Several chapters, especially Chris Uhl's about Penn State and Debra Rowe's about Oakland Community College, explain how to build that support.

What about resources? None of the schools we have described is perilously on the financial edge and none so tight for money that—as in the case of one small African American college in the South—a faculty leader had to wait three years for the budget to stretch to buy recycling bins. Many success stories are connected to grants, budgets, and special funds made available at opportune times. We note, however, that there are cases of universities with financial resources made available and yet sustainability has not taken hold. Clearly, money is not enough and not always necessary.

A number of the institutions described in this book are prestigious and boast strong student bodies and outstanding faculty. Is the sustainability challenge easier in such places? Not necessarily, we conclude. Dick Norgaard, in a personal conversation, notes, "There is tremendous pressure for departments in every field at Berkeley to be among the top three in the country, certainly the top ten, making it difficult to spare resources for innovative interdisciplinary efforts that are not ranked." He concludes that the success of the Energy and Resources Group at Berkeley depended a great deal on luck and on "a touch of mystery."

Spontaneity and persistence really are necessary. Spontaneity—taking advantage of opportunity—is clearly characteristic of successful efforts. Debra Rowe recounts how she worked with a new chancellor with a futures analysis agenda and was able to move ahead on the effort to add sustainability as a universal academic requirement. Change opportunities, in money, moral support, or the teachable moment, are an important tool. Getting the word out takes many forms—news stories, rituals, woods walks, report unveilings—and several chapters teach important lessons in using events as platforms for reaching beyond the inner group of active participants. Persistence is a key theme is Chris Uhl's account of change at Penn State—and we honor Mike Edelstein's ability to make lemonade from lemons, as Ramapo College takes "one step forward and two steps back." In several cases, patience, persistence, and resilience lead to better long-term outcomes despite short-term setbacks.

Emerging Dimensions for Higher Education

The new institutional forms that result from the efforts recounted in this book remind us that "we bequeath social institutions, as well as an 'environment' to future generations" (Redclift 1997, 266). These chapters describe changes in routines, policy, procedures, mission, identity—for individuals, units, and whole institutions. More specifically, these transformations have led to a variety of new forms and commitments:

- *New institutional units* Emerging on several campuses around the country are environmental officer/cordinator positions that bring a budget, staff, and authority to address environmental concerns. Councils, such as Indiana University's Council on Environmental Stewardship, can be highly effective in promoting change.
- *New institutional resources* Success at Middlebury College led the environmental studies area there to receive the status of "center of excellence," thereby signaling it to be one of six units to receive special resources.
- *New clout with donors* Several universities have found that environmentally sensitive green buildings or other environmental initiatives garner attention from alumni and other private donors.
- *New partnerships with federal agencies* The Michigan State University is an example of campus collaboration with the U.S. Environmental Protection Agency.
- *New faculty development programs* Both the Northern Arizona University and Emory experiences with curriculum greening led to new professional research directions, innovative teaching methods, stronger intellectual ties across schools, and deeper faculty satisfaction with the institution (with a positive impact on retention).
- *Transforming the intellectual mission* Transforming the content of education to include sustainability issues takes many forms, including changes in the methods of education as well. Hands-on classes, research projects, and clubs offer important venues for new perspectives. Also useful are new courses, new all-campus requirements, modules and exercises in existing courses, and opportunities for linkage with K–12 education. Some of the most creative examples involve linkages among these types of teaching. These efforts also lead to new questions: What is environmental literacy, not only for students, but for administrators, faculty, and staff as well?

Though one might label each change listed above as narrow or broad, our histories show it is hard to assess effectiveness in the short run. A failed effort to shift a small unit at one point becomes a key moment in a larger transformation later. Our stories also suggest that sustainability efforts can profitably emerge at different levels of social process, such as:

- Faculty legislatures for the adoption of curricular reform and policy changes
- Administrative units that choose new guidelines for everyday campus functioning
- Coalitions of students (or multiple stakeholders) that target a particular action

Thus, although some leaders may experience real risks, new venues of power, new arenas for effective education and collaboration, and even new research productivity can develop as a result of efforts such as these.

Restoring the Intrinsic Rewards of the Academy

Our narratives reveal deep satisfactions in sustainability work. The cooperative efforts required for most campus environmental change mark a shift from the individualistic, entrepreneurial, disciplinary culture of many universities, and participants report such face-to-face activities to be deeply gratifying. Sustainability efforts often form a third sector between the formal structures of university governance and the private domain of the classroom and research. They counter the alienation some faculty feel in the bureaucratized and productivity-driven climate of their institution.

Sustainability efforts have the reward of bringing personal actions in line with ethics and values that seek to do less harm to the natural environment. Some leaders are encouraged by the gratitude expressed by others on campus who are relieved that "somebody is finally doing something."

Among the intrinsic satisfactions of sustainability work is an intellectual dynamism that feeds a lively curiosity. Slowly, this work seems to restore a sense of community—and for some, even a sense of wholeness—that was missing in many institutions of higher education. Campus activism "may be exhilarating and provide one with a sense that one makes a difference in creating a more just world" (Baer 2002, 48). There are joys of building attachment to the natural world, sharing that attachment with others, and creating "sacred places" on campus. All of these intrinsic satisfactions of working toward campus sustainability are a powerful force to counter exhaustion, burnout, discouragement, and despair. Although the momentum of industrial society is enormous, the positive rewards are very real in the lives of campus leaders and allow us to sustain ourselves in this work.

Over and over, our stories emphasize the power of simply bringing committed and curious people together in one room. They speak to allowing wisdom to emerge from a group and of using the energy people bring to build new, creative syntheses of ideas, programs, or policies. In a very real way, what we have learned is that it is the willingness to listen to stories that makes the difference and to "recall or tell a portion, if only a small detail, from a narrative account or story." As Leslie Silko notes in the epigraph to this Introduction, such telling and retelling enable a community to "piece together valuable accounts and crucial information that might otherwise have died with an individual." Indeed, as one of Barry Lopez', characters tells us, "Sometimes a person needs a story more than food to stay alive" (1998, 59). The way forward in the transformation of higher education is not always clear, nor the steps easy, but hearing the voices of those who have gone before enables us to create new stories of how we might envision a sustainable world. Widening the circles of involvement, change unfolds, and the future stories are written.

References

Aronowitz, Stanley. 2001. *The Last Good Job in America: Work and Education in the New Global Technoculture.* New York: Rowman and Littlefield.

Baer, Hans A. 2002. "On the Nature of Our Workplace as Academic Anthropologists: Implications for Critical Teaching and Research." *Practicing Anthropology* 24(3):46–48.

Barth, Fredrick. 1994. "A Personal View of Present Tasks and Priorities in Cultural and Social Anthropology." In *Assessing Cultural Anthropology* (pp. 349–361). Edited by Robert Borofsky. New York: McGraw-Hill.

Boyer, Ernest L.1990. *Scholarship Reconsidered: Priorities of the Professoriate.* Princeton, N.J.: Carnegie Foundation for the Advancement of Teaching.

Brundtland, Gro, ed. 1987. *Our Common Future.* New York. Oxford University Press and World Commission on Environment and Development.

Clugston, Richard M., and Wynn Calder. 1999. "Critical Dimensions of Sustainability in Higher Education." In *Sustainability and University Life* (pp. 31–46). Edited by Walter Leal Filho. Frankfurt: Peter Lang.

Dewey, John. 1916. *Democracy and Education: An Introduction to the Philosophy of Education.* New York: Macmillan.

Fischer, Frank, and Maarten A. Hajer, ed. 1995. *Living with Nature: Environmental Politics as Cultural Discourse.* New York: Oxford University Press.

Geller, E. Scott. 1994. "The Human Element in Integrated Environmental Management." In *Implementing Integrated Environmental Management* (pp. 5–26). Edited by J. Cairns, Jr., T. V. Crawford, and H. Salwasser. Blacksburg, Va,: Virginia Polytechnic Institute and State University.

Graham, Hugh Davis, and Nancy Diamond. 1997. *The Rise of American Research Universities.* Cambridge, Mass.: MIT Press.

Grant, Gerald, and David Riesman. 1978. *The Perpetual Dream: Reform and Experiment in the American College.* Chicago: University of Chicago Press.

Holmberg, Johan, and Richard Sandbrook. 1992. "Sustainable Development: What Is to Be Done?" In *Making Development Sustainable: Redefining Institutions, Policy, and Economics* (pp. 19–38). Edited by Johan Holmberg. Washington, D.C.: Island Press.

Kimball, Bruce. 1986. *Orators and Philosophers: A History of the Idea of Liberal Education.* New York: Teachers College, Columbia University.

Leal Filho, Walter, ed. 2002. *Teaching Sustainability at Universities.* Frankfurt: Peter Lang.

Lopez, Barry. 1998. *Crow and Weasel.* New York: Farrar, Straus and Giroux.

Meiklejohn, Alexander. 1981. *The Experimental College.* Edited and abridged by John Walker Powell. Cabin John, Md.: Seven Locks Press.

National Wildlife Federation. 2001. *State of the Campus Environment: A National Report Card on Environmental Performance and Sustainability in Higher Education.* Washington, D.C.: National Wildlife Federation.

Orr, David W. 1993. *Earth in Mind: On Education, Environment, and the Human Prospect.* Washington, D.C.: Island Press.

Redclift, Michael. 1997. "Postscript: Sustainable Development in the Twenty-first Century: The Beginning of History?" In *The Politics of Sustainable Development* (pp. 259–268). Edited by Susan Baker et al. New York: Routledge.

Rogers, Everett. 1971. *The Communication of Innovations: A Cross-Cultural Approach.* 2nd edition. New York: Free Press.

Roseberry, William. 1994. *Anthropologies and Histories: Essays in Culture, History, and Political Economy.* New Brunswick, N.J.: Rutgers University Press.

Sabin, Paul. 2002. "Academe Subverts Young Scholars' Civic Orientation." *Chronicle of Higher Education,* February 8.

Sharp, Leith. 2002. "Green Campuses: The Road from Little Victories to Systemic Transformation." *International Journal of Sustainability in Higher Education* 3(2):128–145.

Silko, Leslie Marmon. 1996. "Landscape, History and Pueblo Imagination." In *The Ecocriticism Reader: Landmarks in Literary Ecology* (pp. 264–275). Edited by Cheryll Glotfelty and Harold Fromm. Athens: University of Georgia Press.

Trzyna, Thaddeus C., ed. 1995. *A Sustainable World: Defining and Measuring Sustainable Development.* Sacramento and Claremont, Calif.: International Center for the Environment and Public Policy.

Veysey, Lawrence R. 1965. *The Emergence of the American University.* Chicago: University of Chicago Press.

Viederman, Stephen. 1995. "Knowledge for Sustainable Development: What Do We Need to Know?" In *A Sustainable World: Defining and Measuring Sustainable Development* (pp. 36–43). Edited by Thaddeus C. Trzyna. Sacramento and Claremont, Calif.: International Center for the Environment and Public Policy.

Wilk, Richard R, ed. 1989. *The Household Economy: Reconsidering the Domestic Mode of Production.* Boulder, Colo.: Westview Press.

I

Laying the Cornerstones

1

Process and Practice: Creating the Sustainable University

Christopher Uhl

The Pennsylvania State University is situated in a fertile limestone valley, surrounded by forest-covered sandstone ridges. The main campus, covering almost 300 acres, is located in State College, a town of about 60,000. PSU is the land grant school for the commonwealth of Pennsylvania and has an enrollment of 34,500 undergraduate students and 6,300 graduate students.

When I began teaching environmental science at Penn State in 1982, I imagined that the environmental problems that I was teaching about were "out there" in the "real" world and had little to do with the day-to-day operations of my university. Indeed, because universities are powerhouses of knowledge and expertise, I assumed that they would be solving our environmental problems and modeling sustainable practices. Even if they were not, I was too busy with "important" research to pay attention to something as mundane as the day-to-day physical operations of my university.

My research at that time (1980s through mid-1990s) was centered on the human activities leading to the biotic impoverishment of Amazonian ecosystems. Then (and, lamentably, still today) humans were aggressively extracting Amazonia's riches. Miners were digging up gold and bauxite, loggers were scouring the forest in search of high-value hardwoods, fishermen were depleting the rivers of fishes, and farmers and ranchers were replacing the verdant forest with cassava fields and weedy pastures. Little of what I saw in Amazonia was sustainable.

In the evening, I would often hang out with Brazilian friends, and we would sometimes discuss the myriad threats to the rain forest. One night when I was feeling particularly despondent, Ana Cristina said, "Hey, things aren't so bad here, my friend. At least we still have 75 percent of our forest intact. You guys in the States have already cut 95 percent of your primeval forest, and now you are hacking down the last few percent in the Pacific Northwest." Of course, she was right.

Later that night, I went to a movie by myself. The film was *Pretty Woman* (the movie houses along the Amazon usually show popular Hollywood flicks). I decided to watch the movie not as a lonesome American but, instead, imagining I was a native of Amazonia. Hence, what I saw depicted on the screen was not the little love story featuring Julia Roberts and Richard Gere, but instead the glorification of a whole way of life based on materialism, speed, and shallow relationships—all packaged in a way to make it seem fun and glitzy. Suddenly, the United States wasn't a country but a "brand" that was being marketed to the world. I left the theater knowing more clearly than I had known before that the American approach to life—based as it so often is on money, acquisition, and instant gratification—is colonizing the psyches of the world's people. The United States is the model and right now its compass points the entire world toward a nonsustainable future. But the United States could be leading the way to creating a sustainable world. Furthermore, U.S. universities, as centers of innovation and learning, could be in the forefront, leading the charge.

Eventually, I decided to shift my attention from distant and exotic Amazon ecosystems to the seemingly ordinary ecosystem right in front of my nose: Penn State University. I reasoned that a necessary first step to encourage sustainability at Penn State would be to take a baseline measure of university operations, with an eye to ecological performance. Although I did not foresee it at the time, this early work would attract other faculty members as well as students and lead to the formation of a research team, and this team would develop indicators that would reveal the degree to which the university was moving toward or away from sustainable practices. Once our team had pinpointed where the university stood, we were positioned to articulate a clear vision for where the university needed to go to become ecologically sustainable. This understanding prompted us to develop strategies to incorporate this vision into an ecological mission for the university. The final step, which continues to occupy us, is to translate the university's newly adopted ecological mission into concrete policies and actions.

Our experience at Penn State illustrates this three-step process of developing sustainability indicators, then an ecological mission, and finally policies to institutionalize sustainable practices.

Developing Sustainability Indicators

As I was leaving the biology building late one winter evening in 1996, I looked up and saw lights on in many of the labs. Biologists often get their best work done in the still of the night. Often they work alone. I too was accustomed to doing research alone, but I wanted this new research initiative on sustainability to have a more open and inclusive quality about it. I believed that the research *process* would be as important as any final research paper or report. And I knew from the start that the results of the research were not so much intended for scientific journals as they were for the students, staff, and faculty of Penn State and other universities.

I inaugurated the new initiative by posting an announcement on a bulletin board in the Penn State Student Union, inviting students to participate in a study of the "ecological sustainability of Penn State." Nine students expressed an interest in the project, and we met to hatch a plan for measuring sustainability. I was candid with the students, telling them that although I knew how to measure the dissolved oxygen concentration of a lake and the acidity of soil, I did not know how to measure sustainability. Indeed, there is no equipment manufacturer that sells a "sustainability meter."

In an effort to invite the students into the problem, I asked them to think about Penn State as an ecosystem. In what ways was the university similar to—and in what ways different from—a natural ecosystem? The students observed that in nature, everything cycles. In contrast to natural ecosystems, the flow of materials in human-engineered ecosystems, like Penn State, is mostly linear—one way. Indeed, our universities are constantly receiving materials from distant sources, consuming these materials, and then shunting the wastes to distant "sinks."

The students believed that these linear pathways of material flow were extremely wasteful, and this bothered them. They complained about the way that people at Penn State wasted water, electricity, paper, and food. I invited the group to spend time thinking about how we might measure consumption and waste at Penn State. We continued to meet over the next two months, but then interest began to wane. When I asked why we were losing our momentum, the students made it clear that they were tired of hashing things out; they wanted to take action.

Making the Invisible Visible

We began by looking at the university's underbelly or backside. Both individually and in small groups, students visited the landfill that receives Penn State's trash, journeyed to the open pit mines that provide Penn State's coal, and walked through the well fields supplying the campus with water. They looked into dumpsters to see what Penn State people were throwing away, traced the sources of the food served in university dining halls, studied land transactions at the county deeds office, conducted botanical surveys of the campus grounds, and much more.

Rather than sitting in classrooms and talking about the state of the environment, these students were engaging in face-to-face interactions with Penn State's complex and often invisible support systems and the people responsible for running them. As they conducted their investigations, they realized that many of the ways in which the university relies on the environment are hidden from view. Hence, as a team, we decided to center the first phase of our work around the theme of "making the University's invisible ecological dependencies visible." We thought that a good way to do this would be through personal stories (see the box).

Using Sustainability Indicators

The stories, like Amy's, were a useful starting point for looking at Penn State through the lens of sustainability, but something more comprehensive was needed. It took our team a while to figure out what that would be. One day while I was walking past Old Main at the heart of the Penn State campus, it struck me that universities are like entire societies in miniature—they have their food system, their energy system, their water system, their transportation system, and so forth (figure 1.1). If we could develop markers, or indicators, of sustainability for each of the university's subsystems, then we could gauge the ecological health of the university.

Our team soon discovered that we were not alone in our quest for sustainability indicators. Governments, organizations, and cities around the world are beginning to develop ways of tracking their progress toward sustainability. We were particularly inspired by a report that described how citizens in the city of Seattle had agreed on forty indicators of sustainability <www.sustainableseattle.org>.

As our work became more focused, more people began coming to our meetings and planning sessions. Several dozen Penn Staters participated

Amy's Dorm Room

When Amy Balog was a Penn State junior, she wanted to know how much coal she and the other students in their dorm, Beaver Hall, were consuming each day as they flicked their lights and computers and stereos on. She began knocking on doors and asking fellow students if she could count the number of plug-in devices in their rooms. She found that a typical dorm room had twelve plug-in devices: micro-fridge, television, VCR, computer, printer, alarm clock, CD player/radio, answering machine, video game unit, and several lamps. Some rooms had as many as 19 plug-ins.

Amy then administered a questionnaire to gauge the number of hours that the various plug-ins were in use each day. Next, she used a watt meter to measure the energy consumption for each category of plug-in. Crunching the numbers, she determined that, on average, 10 kilowatts of electricity—or 8 pounds of coal—were used to supply the daily electricity needs of each dorm room. Scaling up to the entire dorm, Amy estimated that a little more than a ton of coal is required to supply Beaver Hall's total electricity needs each day. The burning of this coal releases about 3 tons of the greenhouse gas carbon dioxide to the atmosphere.

As students considered the implications of Amy's findings, they discussed ways of making this invisible connection—between electricity use and fossil fuel consumption—visible. One student suggested that an 8-pound chunk of coal be placed on all dorm room desks and a ton of coal set by the entrance to all dorms.

in defining the sustainability indicators. We began this process by defining best or sustainable practices for each university subsystem. For example, we concluded that a sustainable energy system should be based on renewable energy and be highly efficient and nonpolluting. Hence, our energy indicators measured if Penn State's energy system was becoming less dependent on fossil fuels, less wasteful, and less polluting over time.

In all, we developed thirty-three indicators for gauging sustainability <www.bio.psu.edu/greendestiny>. Guided by these indicators, we scrutinized Penn State's policies and performance in water conservation, recycling, purchasing, landscaping, energy use, building design, and research ethics. We critically evaluated the food and transportation systems and asked if the university was moving in a sustainable direction. We checked to see if Penn State's institutional power was being used to strengthen regional economies and promote corporate responsibility, and much more.

Figure 1.1
Drawing of Old Main showing the various university subsystems

Students did most of the initial work. They picked an indicator that they were interested in and developed a plan of study. Sometimes these were independent study projects undertaken for credit with faculty guidance; sometimes they were part of the content of an environmentally oriented course.

In most cases, the data for the indicators already existed but had never been used to assess sustainability. For example, by studying a sequence of preexisting university maps, a Penn State senior, Nate Hersh, determined that the proportion of green space covered by impervious surfaces on campus increased by 50 percent between 1970 and 2000.

Often the data for the various indicators could be plotted, and, depending on the trends over time, indicated a movement toward or away from sustainability. For example, total waste production increased by over 20 percent at Penn State between 1989 and 1999 (more than two times the increase in the Penn State population for the same period).

Early on in this indicators study, I had a meeting with the university provost to tell him about our project. He listened attentively while I described the various sustainability indicators we were using. When I finished, he expressed support but cautioned against using qualitative indicators, saying that the inclusion of such indicators would compromise the rigor of the work. His words affected me deeply. As a scientist, rigor is important to me. I know that my colleagues are quick to denigrate qualitative inquiry, often characterizing it as soft or fluffy.

It was tempting to follow the provost's counsel and define sustainability in strictly biophysical terms, as many have done. But this would have meant restricting our work to an auditing exercise. In the end, our team decided against this approach because we felt that a significant part of what is important and worthy of attention in life cannot be expressed in numbers. Indeed, sustainability is about much more than millions of Btus saved or tons of paper recycled. It is a heartfelt way of looking at the world that encompasses mindfulness of place, respect for natural processes, discernment of true needs, honesty, and civic responsibility.

By including qualitative indicators, we have been able to raise questions that get at the soul of sustainability. For example, we thought that it was important to pay attention to the effects of technology on sustainability so we created an indicator called "Technology: Enhancing vs. Undermining Community?" In our analysis for this indicator, we provided data but also invited the university community to reflect on technology's problematic aspects (see the box).

The first Penn State Indicators Report, released in 1998, depicted an institution whose performance, measured by sustainability indicators, was not exemplary. In category after category (energy, food, materials, transportation, building, decision making), Penn State practices departed little from the national status quo. The university's official posture appeared to be in accord with the national view that we can continue with business as usual—growing and consuming—without worry. And yet in private conversation, people in all sectors of the university were concerned about the deterioration of the environment worldwide and overconsumption in the United States, in particular.

Using ecological indicators to give the university a report card was unsettling to some Penn State administrators. After all, they did not commission this study, and there was legitimate concern that our findings might tarnish the image of the university. Indeed, we were tempted to assume a highly critical posture because the university's environmental performance was lackluster in many areas. In the end, though, we decided against a highly confrontational posture because we came to see that our goal was not to win arguments but to effect long-term change.

Can Some Technologies Undermine Community?

The choice to adopt a technology to do something that we previously did on our own is not always trivial. Consider Penn State's decision to replace the hand rake with the leaf blower. The leaf blower technology has certain characteristics and affirms certain values. When we use it, we are opting for a fast (machine) pace rather than a natural pace, noise rather than quiet, polluted air rather than clean air, and so forth. These things—fast pace, polluted air, and noise—can have a negative effect on the frequency and quality of our social interactions (i.e., the quality of community life). Leaf blowers are an obvious case, but almost all of the technologies (answering machines, computers, motor vehicles, televisions) that we have adopted over the last century have the potential to affect the quality of our community life for better or worse. So far, we at Penn State have been disinclined to critically examine the possible negative effects of our myriad technologies on the quality of community life <www.bio.psu.edu/greendestiny>.

Nonetheless, sometimes our ardor and insistence on transparency caused problems for us. After all, it takes a good deal of ideological commitment to sustain such an effort, and the same ideological commitment caused us, at least initially, to say what we felt was right, regardless of the political consequences. For example, we made the mistake of sharing the first draft of the report, which did not mince words, with a top administrator. He was clearly perturbed and complained that the report was excessively negative. This created a testy climate that took a long time to overcome. From that point on, we attempted to cite the positive things that the university was doing while also making the university's shortcomings transparent.

We gradually learned that each organization has its own change model, its particular way of changing. At Penn State, significant ideological shifts are effected very slowly. The way to change things is with persistence, not insistence. Showing how problems are actually opportunities creates a dynamic tension that is pregnant with energy and excitement.

As we prepared to release the first Indicators Report we invited university leaders (e.g., deans, department heads, unit heads) to supply written endorsements in an effort to create a positive buzz around the report. The associate dean of liberal arts had this to say: "This report is a demonstration of the kind of exciting and relevant learning that can take place when students and faculty work collaboratively. The sustainability project demanded methodological rigor and an interdisciplinary, integrated systems approach to the problem. But it also required the participants to grapple with ethical and moral questions involving distributional justice and the responsibility of the University to society. Penn State should be proud of the result." These endorsements were included on the front and back covers of the report and in the announcements heralding the report's release.

The report was formally released to the university in a large open-air public ceremony on the steps of Old Main. Copies were sent to all department and unit heads. Leaders from various sectors of the university's Office of Physical Plant (the energy czar, the head of landscaping, the chief of waste management, the transportation coordinator, and others) were on hand to receive copies of the report. They were the unsung heroes of this effort because they and their staff had spent immense amounts of time tracking down data, talking with students, and checking over early drafts of the report for accuracy.

After the report's release, some faculty members from across the university—in agriculture, engineering, landscape architecture, ecology, political science, and communications—voluntarily began to use the entire report or parts of it to teach about sustainable practices, environmental ethics, place-based research, rhetoric, citizenship, and so forth.

An important general lesson of this sustainability indicators work is that institutions measure only what is important to them. And there is nothing more important for humanity's future than moving forthrightly from practices that harm the earth to practices that are sustainable. This means it is time to measure sustainability not just in universities, but in all realms of society—government, business, education, religious institutions. Sustainability is a whole new way of seeing and relating to the world, and the act of measuring it legitimizes it.

Our sustainability group experienced a sense of satisfaction in the fall of 1998 after releasing the Indicators Report. We were in the news. Reporters were calling us from all over the East. Pennsylvania's Department of Environmental Protection was requesting a box full of the reports to distribute to their personnel, and students and faculty from dozens of universities were contacting us to request copies of the report. Penn State's president asked that a copy of the report be sent to all members of the board of trustees, and he was passing the report on to his vice presidents, instructing them to study its recommendations. With all this activity, it was tempting to imagine that our work was finished. After all, the report clearly documented the gaping sustainability deficit at Penn State and prescribed thirty concrete steps that Penn State needed to take to erase this deficit.

But six months after the report's release, very little had ostensibly changed. Reluctantly, we acknowledged that the Indicators Report, by itself did not have the power to transform Penn State into a sustainable university. Nevertheless, it provided the language to begin to talk about sustainable practices at Penn State. As with any other attempt to change the status quo, persistence would be essential.

Up to this point, we were just a couple dozen university folks (mostly students) who had come together around a common concern. We eschewed formal membership, a constitution, rules, or official university standing and in this way avoided many of the problems that institutionalization and bureaucratization might have created. It was our allegiance to sustainability and our desire to transform PSU to "Pennsylvania's Sustainable University" that united us. Although our internal structure was very open and informal, we did establish a Web site, and when the occasion demanded, we were ready to portray ourselves with formality.

We also spent a long time coming up with a name for ourselves. Names matter a lot. When the folks in Seattle hit on "Sustainable Seattle" for their fledgling group, they must have known that they had a winner: the name of their town plus the name of their mission, linked by alliteration.

After trying out lots of possibilities for our group, we finally hit on "The Green Destiny Council." This name was inspired by Penn State's multiyear $1 billion fund-raising effort dubbed "Grand Destiny." By substituting the word *green* for *grand*, we signaled that ours was a group concerned with ecology and the environment; by playing off "Grand Destiny," we had a name that people would remember (especially decision makers); and by using the word *council*, we conveyed the egalitarian character of our organization.

One year after the release of the first Indicators Report, we made a commitment to release an updated and expanded version of the report in the year 2000. This allowed us to keep the university's environmental performance in the spotlight.

Developing an Ecological Mission for the University

In the period following the release of the first Indicators Report in 1998, the big question before our group was, "What's next?" Toward the end of one of our Friday afternoon meetings, a faculty member said, "What we really need to do is institutionalize sustainability." A student asked, "How would we do that?" After a long silence, the faculty member responded, "We could do it by making sustainability central to Penn State's mission." Immediately, there was ripple of excitement; this was an idea that offered us traction.

A small group (myself and two students) spent three months drafting Penn State's ecological mission. On the face of it, this seemed ludicrous—two students and a professor drafting the university's ecological mission:

we had no vested authority to do this. But we had learned that we did not need to wait for permission; we could just begin the process.

We called the mission document, "Green Destiny: Penn State's Emerging Ecological Mission" <www.bio.psu.edu/greendestiny> to signal that we were working as midwives to birth a mission for the University. Each of the document's eight core pages proposed a facet of the new ecological mission (see the box).

We knew that it wouldn't work for us simply to declare what we thought the university's ecological mission ought to be. We would have to open up the process and cultivate support, especially among faculty and staff in positions of leadership. In other words, we would have to schmooze.

I began the schmoozing process with personal phone calls to every department head, dean, assistant dean, unit head, and facilities chief on campus—almost 150 leaders. The conversations typically went something like this:

"Hi, Joe. This is Chris Uhl over in Biology."

"Hi, Chris."

"Listen, Joe, I don't think we have met, but I wonder if I could ask your help with something. It has to do with Penn State."

"Sure. What is it?"

"Well, I have been working with a group called Green Destiny Council—you know the folks that released the Penn State Indicators Report a while back."

Green Destiny's "Emerging Ecological Mission" for Penn State

Energy:	Move Toward Fossil Fuel Independence
Water:	End Water Waste
Materials:	Become a Zero-Waste University
Food:	Eat Foods Produced Sustainably
Land:	Create and Abide by a Land Ethic
Transportation:	Promote Alternatives to Car Transit
Built Environment:	Create "Green" Buildings
Community:	Guarantee Ecological Literacy

"Yeah, right. I recall hearing something about that."

"Well, as a follow-up, Green Destiny has put together a much shorter document that attempts to lay out an ecological mission for Penn State. Do you follow?"

"Yeah, I'm with you."

"Joe, I have never been involved in drafting a mission, and this is where I need your help. I wonder if you would look over what we have put together and perhaps comment on it?"

"Sure, Chris. Send it over."

The mission document that the 150 leaders (including all top administrators) received was eye-catching. There was a cover letter with a formal Green Destiny letterhead, and the cover of the document had a color photograph of the Earth along with the Penn State official logo, and a red silk ribbon. On the last page, we asked reviewers to place a check next to each mission element indicating their stance: "support," "don't support," or "undecided." We also encouraged reviewers to include specific reactions to any or all of the mission components.

Support ran high (over 70 percent) for all eight of the mission elements. The second most frequent response was "undecided." The "don't support" response was less than 10 percent in all cases. We modified the language to address what we judged to be legitimate concerns and then summarized the results and sent a short report back to all the leaders. Then we called a meeting with the provost. He expressed genuine support for Green Destiny's mission document and encouraged us to take it to the faculty senate for endorsement.

Meanwhile, the Office of the Physical Plant issued a fifteen-page, generally positive, critique of the Green Destiny's ecological mission proposal, and Penn State's president was beginning to mention sustainability in public. It was also at about this time that *Penn State Research,* a university publication that is sent out to approximately fifty thousand alumni, carried an article about Green Destiny's Sustainability Indicators initiative.

After spending six months in committee and undergoing minor language modifications, Green Destiny's Ecological Mission statement was put to a vote before Penn State's faculty senate and approved unanimously. Next, it went to the president's desk. He quickly added his approval.

After four years of persistence, Penn State now had a comprehensive set of sustainability indicators telling it where it stood and an ecological mission telling it where it needed to go.

After the faculty senate and the Penn State president endorsed Green Destiny's Ecological Mission proposal, we again asked ourselves, "What's next?" It seemed that the time had come to figure out a way to put the lofty ideals and good intentions embodied in Penn State's ecological mission into concrete actions. Specifically, we asked ourselves, "How could we create a detailed blueprint for sustainable practices at Penn State?"

Sustainable Practices: The Mueller Report

Blueprint work is nuts-and-bolts technical stuff; it concerns heating and cooling systems, the design of urinals, the margin settings on printers, the volatile organic compounds in paints, and so forth. One afternoon when we were discussing this, a faculty member said, "These details are pretty boring, but if it was my own house, I'd be interested." We were sitting in the Penn State Biology building, Mueller Lab, at the time. Suddenly I realized that we could create a sustainability blueprint for the very building that we were in.

At the time of these discussions (September 2000), I was teaching a five-credit ecology course in the biology building. It had been my custom to devote the last six weeks of this course to what I called "the ecology in action" project. Instantly I knew I had my action project for the semester. I would give these biology students, with their concern for the complexity and intricacy of life systems, the opportunity to join their knowledge of life with actions in their "home" building that respect and nurture life.

When it came time to initiate this project in early November, I told the twenty students in the class that their assignment was to cut the ecological impact of the Mueller building in half while creating healthier working conditions for all Mueller occupants."

Students began by considering all the inputs to the building: electricity, steam, paper, computers, printers, toners, furniture, carpeting, paints, cleaners, pesticides, coffee, and so forth. Each student took one input and determined (1) Mueller's annual consumption for that item, (2) the environmental impacts of this consumption, and (3) alternatives that would significantly reduce ecological impacts.

They set to work examining the records in the Mueller purchasing department, conducting inventories of the computers and printers in the building, characterizing the floor coverings and the lighting technologies, interviewing the janitorial staff, and so on. They also searched the library and the Web for examples of ecologically benign approaches to carpeting, computing, paper production, and so forth. On the final day of class, they presented their findings to representatives from Mueller, as well as staff from the university's Office of the Physical Plant. Although the students were not able to do an exhaustive analysis, they did a fine job of gathering data and presenting preliminary results.

Next, a new team composed of four recent Penn State graduates, a Ph.D. graduate student in engineering, and myself went to work fleshing out the analysis. Five months later, we had a solid document, which we entitled, The Mueller Report: Going Beyond Sustainability Indicators to Sustainability Action." This report <www.bio.psu.edu/greendestiny> offered the university a blueprint for halving the ecological impacts of its current building stock. The box provides an abbreviated excerpt (stripped of accompanying tables, calculations, and footnotes) that captures a taste of the report's breadth and analytical approach.

In the process of conducting the Mueller study, we learned that the lion's share of the building's ecological footprint was in energy consumption. Indeed, this building requires more than 2,200 tons of coal per year for its operations, the burning of which releases over 5,750 tons of carbon dioxide. On a per capita basis, the numbers are sobering: 18 tons of coal and 47 tons of carbon dioxide per person (123 building residents) per year. We determined that Mueller's energy consumption could be reduced by half—for example, by switching to energy-efficient computers, printers, and lighting fixtures and by subjecting Mueller's heating, ventilation, and air-conditioning system to a comprehensive tune-up. These changes would save approximately $50,000 annually. When scaled to the entire university, potential cash savings from Mueller-style energy-efficiency retrofit are in the vicinity of $10 million. <www.bio.psu.edu/greendestiny>.

In addition to energy analyses, we detailed ways of significantly reducing Mueller's waste associated with the use of water, transparencies, diskettes, printer cartridges, computers, carpeting, and furniture. We also drafted model policies for all Mueller materials. For example, the proposed carpet policy reads as follows:

Mueller Paper

The 123 faculty and staff occupying the Mueller Building consume, collectively, 5.3 tons of chlorine-bleached, 0 percent post-consumer-content paper each year. Mueller's paper comes from Willamette Industry's paper plant in Johnsonburg, PA. In 1998 that plant released 338 tons of pollutants, including 61 tons of sulfuric acid and 148 tons of hydrochloric acid.

Mueller could significantly reduce its paper "footprint," first, by purchasing 100 percent post-consumer-content paper that is chlorine free; and, second, by more fully utilizing the paper that it purchases. At present, Mueller documents are often printed without considering how font size, margin width, and line-spacing decisions affect paper needs. Paying attention to these "details" can dramatically reduce paper consumption. For example, a hundred-page "standard" print job (i.e., 12-point font, standard margins, double spaced, one-sided) can easily be reduced to less than 20 pages by reducing font size to 10-point, extending top, bottom, and side margins to 0.75", and using single spacing and 2-sided printing.

By buying 100 percent post-consumer recycled paper and fully using that paper, Mueller could reduce its annual paper use by two-thirds, from just over 1 million sheets to approximately 300,000 sheets. Expressed on a per capita basis, a Mueller occupant adopting "best" paper practices would decrease his/her paper consumption from over 8,000 to approximately 2,700 sheets, and, in so doing, save over 555 gallons of water, about 360 kWh of electricity, approximately 2,650 square feet of forest land, and almost 800 pounds of CO_2 emissions. Moreover, although recycled paper costs more per sheet, the potential reduction in paper use could reduce per capita paper expenditures by $25 per year.

Adopting even the most simple paper conserving strategies at the scale of the entire University could result in significant monetary savings. For example, if Penn State was to change standard computer/printer margin settings to 0.75" on all sides (making 19 percent more area available on each text page), the University would reduce annual paper consumption by 45,000 reams and save $123,000 each year <www.bio.psu.edu/green destiny>.

Mueller Laboratory, through its strong commitment to environmental stewardship, seeks to reduce the environmental impact of its carpet use. In order to accomplish this objective, the following steps will be taken during the procurement and disposal of carpeting:

- Give preference to pre-existing tile rather than carpet.
- Purchase carpets having 100 percent post-consumer recycled content and solution or vegetable dyed fibers.
- Purchase modular, as opposed to broadloom, carpet to the extent that the quality and end-use of the floor covering remains uncompromised.

- Purchase carpets and adhesives having the lowest VOC level available.
- Lease carpet from Interface Inc. or a similar company, or send old carpet to a recycling center.

Detailed policies like this are essential for creating a sustainability blueprint. Indeed, policies are what give an ecological mission its traction.

Although the Mueller Report was ostensibly about how to reduce the ecological impacts of the university's campus building stock, the broader message was that the campus buildings squander massive amounts of energy and money. These buildings were constructed at a time when most people imagined that U.S. supplies of energy were nearly inexhaustible and almost no one had made the connection between fossil fuel use and climate disruption. We live in a different time. We know much more, which means that we need to do much more. By employing green design technologies, it is now possible to achieve eight- to ten-fold reductions in energy use. For example, the Commonwealth of Pennsylvania has just completed an office building in Cambria County that uses only one-eighth as much energy per square foot for heating and cooling as the Mueller building requires.

Prior to the release of the Mueller Report, we asked twenty respected university leaders to review and comment on it. All endorsed the report with enthusiasm. A professor from landscape architecture had this to say: "My hope for this report is that it's read from cover to cover by all Penn State students, faculty and administrators. Why? Because so many of us learn, work and live in wasteful, ugly and in many ways 'unwell' environments. With meticulous investigation and spirited reason, this report shows how a single, rather mundane building—and an entire campus—can be revitalized for the 21st century."

In October 2001, Green Destiny Council released the Mueller Report to the university in a public ceremony. University officials from the Office of the Physical Plant, who had played a key role in providing and interpreting data, were on hand to formally receive the report.

After the report's release, we moved quickly to set up meetings with key decision makers (e.g., the chair of biology, vice president for business and finance, head of university operations). Receptivity was high. Everyone likes win-win situations, and the report was being seen in this light. The Office of the Physical Plant announced its readiness to institute the suite of energy recommendations necessary to reduce Mueller's energy consumption dramatically.

During this same period (2001) and in part as a result of Green Destiny's efforts, Penn State released its first Environmental Stewardship Strategy. As noted on the university Web page <www.psu.edu/oldmain/fab/dstrat/strategy8.htm>, "The Environmental Stewardship Strategy was created to identify specific actions and objectives aimed at conducting the University's business in a manner that demonstrates a commitment to environmental stewardship." The strategy articulates principles of environmental stewardship in the realm of (1) responsible purchasing, (2) efficient use and conservation of energy, water, and other resources, (3) minimization of solid waste production, (4) minimization of hazardous and toxic materials on campus, and (5) environmentally responsible campus design. For example, regarding responsible purchasing, the strategy commits to making environmentally and fiscally responsible purchasing choices that consider life cycle costs, energy use, and long-term disposal implications. To this end, the strategy "encourages obtaining goods that minimize waste products, have high recycled content, use environmental production methodologies, demonstrate maximum durability or biodegradability, repairability, energy-efficiency, non-toxicity, and recyclability."

The strategy contains specific actions that the university is now taking within designated time frames:

- Join the Energy Star Buildings Program by March 2001 (completed).
- Acquire and evaluate the use of waterless urinals by July 2002 (completed).
- Evaluate the purchase of a portion of electric load from renewable energy sources by July 2002 (completed).
- Identify products that can be returned to the manufacturer at the end of their useful life for reuse or recycling by July 2002 (completed).
- Develop or Integrated Pest Management policy by July 2001 (completed).
- Design new facilities using Leadership in Energy and Environmental Design (LEED) criteria to achieve LEED certification of every major campus project (in process).

At long last, Penn State is beginning to operationalize sustainable practices. It is a small but important beginning. Our Green Destiny Council will continue to raise the bar . . . with persistence, not insistence.

Conclusion

Over the years that I have been working on sustainability issues, I have come to understand that sustainability is a social change movement. In this context, Green Destiny's work has really been about alerting Penn State to a problem, as well as an opportunity, and encouraging the university on to a new path. Our success, to the extent that we have had any, has been hinged to our understanding of power and the process of social change and our use of an array of tools and strategies.

As with any other change movement, we have met resistance. At first, the university's administrators assured us that Penn State was already "doing all this environmental stuff"—in other words, everything was under control, and we did not need to worry. This is the way most institutions respond to the prospect of change.

Given the culture of our institution, we needed numbers, indicators, and benchmarks to begin the awakening process. As is true of all social change movements, we also needed trigger events to heighten awareness about the problem and the opportunities. The fanfare we were able to create around the public release of our various reports has served this function.

Now, after five years of persistent effort, it appears that the Penn State population and administration recognize the importance of instituting sustainable practices. Indeed, I smiled when I received a recent note from our president in which he wrote, "I appreciate your efforts to enhance Penn State's sustainability efforts." What I especially liked about this sentence was not the president's sentiment of gratitude but his phrasing: "Penn State's sustainability efforts." You know you are making progress in a social change movement when the target of your efforts begins to assume ownership of the very goals and ideals you have been endeavoring to promote.

2

The Green Task Force: Facing the Challenges to Environmental Stewardship at a Small Liberal Arts College

Abigail R. Jahiel and R. Given Harper

Illinois Wesleyan University is a private liberal arts college of 2,100 students, located amid the corn and soybean fields of central Illinois. The 70-acre campus, situated in downtown Bloomington, Illinois, lies midway between Chicago and St. Louis and draws its students predominantly from surrounding rural communities and the Chicago metropolitan area.

In spring 2000, Illinois Wesleyan University (IWU) created the Green Task Force (GTF), comprising approximately fifty students, faculty, staff and administrators. Its mission was to propose ways to reduce the university's environmental footprint, and we were to direct this effort. The GTF's accomplishments during its brief two-year history were impressive. It established a significantly expanded recycling program and systematized a dumpster dive procedure in which all campus wastes are collected and the rate of recycling for the campus is calculated. The GTF also increased environmental awareness, oversaw a campus energy audit, and started to reduce campus paper and electricity consumption. Not only did the university proudly acclaim these accomplishments, but this work was recognized by a local environmental organization and showcased on the National Wildlife Federation's Campus Ecology Web page.

These achievements did not come easily. Support was weak from those whom we had counted on, and ultimately came from where we least anticipated it. Nor is IWU's long-term environmental stewardship ensured. Still, our story provides insights and hope for small liberal arts colleges like ours, without a history of environmental concern or an activist student body. To understand the challenges we have faced, we begin in 1991, the year Given joined the faculty.

Planting the Seeds

Given: When I first came to Illinois Wesleyan University, I was impressed by the beauty of the campus—the many large trees on the quad, the well-maintained buildings, the carefully manicured lawns and green athletic fields. I was also struck by the lack of recycling bins. A few years later, after being hounded by one student, the university initiated a small-scale recycling program under the leadership of a student coordinator and purchased a few bins. However, the bins were deliberately placed in areas where they were not highly visible. This campus policy of restricting bins in public places was openly challenged only once, when an individual defiantly spray-painted a recycling bin gold to make it "aesthetically pleasing" and placed it smack in the middle of the brand-new science center! This was an unusual act on our generally conservative midwestern campus.

As an avian ecologist passionate about our responsibility to protect the environment, I felt there was little I could do to sway the administration to pay more attention to environmental matters. Instead, I focused on generating environmental awareness in my students by teaching a course on environmental issues and taking students on field studies of exotic ecosystems such as the Great Barrier Reef and the Costa Rican rain forest. By the mid-1990s, the student recycling coordinator position was terminated, and the recycling program languished. Fortunately, several events soon changed things.

In 1998, Illinois Wesleyan formally established a minor program in environmental studies (ES). Shortly afterward, the ES program was awarded a sizable grant from the Rockefeller Foundation, and a faculty position was added. Abigail Jahiel was hired, and she began to codirect the program with me. These events ultimately helped spawn our campus greening movement.

Abigail: I came to Illinois Wesleyan in 1999, eager to develop the new ES curriculum. As a political scientist, I had spent several years studying the Chinese environmental policy process and was concerned with global environmental conditions. My fieldwork experiences had convinced me that before I could comfortably—let alone convincingly—suggest to the Chinese that their path of development was ecologically unsustainable, I personally had to challenge our own overly consumptive lifestyle in the United States. Americans account for 5 percent of the global population but nevertheless consume about 30 percent of global energy resources, whereas Chinese, with 22 percent of the world population, consume about 8 percent of global energy resources. Yet though I had worked to foster an understanding of environmental equity in my students, I had not seen how I, as an academic, could effect significant changes in practice.

Shortly after I was hired, several colleagues and I attended two Project Kaleidoscope (PKAL) conferences designed to promote the development of ES programs. These conferences convinced us that not only could academics change student behavior but also, if we were going to help build a successful ES program, it was our job to change the university's behavior.

Throughout the fall and early spring of 1999–2000, we discussed how we might begin to green our campus. With encouragement from the associate dean of faculty, I began to plan an experiential learning course for May term 2000 to investigate these matters.

Greening the Campus

Calling for Action

Unforeseen events quickly propelled our efforts. In February, a new student senate vice president was elected. By April, she and her senate colleagues had drafted and unanimously approved a detailed resolution calling on the university to consider comprehensively how its actions had impacts on the environment. We crafted a similar resolution and brought it before the faculty, where it too received unanimous approval. These joint calls from students and faculty prompted the provost to ask us, as directors of the ES program, to set up a campus environmental task force. The work of the May term 2000 class would provide the empirical basis for the new Green Task Force.

Providing the Groundwork

During May, the sixteen students enrolled in the experiential learning course, "Greening the Campus," worked intensely to conduct an assessment of the environmental impacts of five aspects of campus life: solid waste management, energy use, water consumption, grounds management, and dining services. They presented their findings and policy recommendations at a briefing conference attended by top administrators, staff, and others. The president was so impressed by the students' presentation that he asked for fifty copies of the briefing book <http://titan.iwu.edu/~environ/greening2.html> to distribute to members of the board of trustees. The dean of students urged us to convey the message widely to Illinois Wesleyan students.

Building the Green Task Force

We met with the student senate president and vice president in early August to construct the GTF. In devising the organizational structure, we had several considerations in mind. We wanted to make sure that the focus of the GTF would be broad and comprehensive, that it represented a cross-section of campus (students, staff, faculty, and administrators), that leadership was shared with students, and that physical plant personnel were well represented. It was difficult to get students to recognize the value of including physical plant personnel. We had to mandate their inclusion in leadership positions, even at the risk that we as faculty might appear to dominate the decision-making process.

In the end, we established seven committees overseen by a standing committee: Dining Services, Energy Consumption, Environmental Education, Purchasing, Reduce/Reuse/Recycle, Toxics and Grounds Management, and Water Use. The committee chairs and cochairs included two physical plant members, three faculty, and six students. The standing committee was composed of fourteen people, including the heads of each committee, the dean of students, the associate dean of faculty, and the director of physical plant. Our open call for GTF volunteers yielded only a handful of faculty and some students, but surprisingly strong interest from secretaries, physical plant personnel, librarians, staff from other areas of campus, and low-level administrators. The GTF ultimately included managers of custodial services, labor crew, maintenance, and grounds management. In addition, the assistant to the president and the associate director for development volunteered. The structure and composition of the GTF would provide unforeseen challenges as well as unexpected opportunities.

Maintaining the Momentum

The first semester of GTF work was difficult. It quickly became obvious that one committee was thriving while most of the others were floundering. The Reduce/Reuse/Recycle (RRR) Committee met almost every week, while each of the other committees met at most three times all semester. During these meetings, many issues were rehashed, but few strategies emerged. The problem was most severe in the committees chaired by students and staff. The initial enthusiasm expressed by the student senate leaders was not translated into action. Their committees met infrequently, and they attended fewer and fewer standing committee meetings, as other campus issues and postgraduation plans took precedence. Students in general did not have all of the organizational skills required for leadership. Significantly, neither students nor staff were comfortable asserting themselves in front of faculty. Clearly, the traditional hierarchy and transient nature of the university population were impeding the success of our plans.

In our biweekly standing committee meetings, we tried encouraging committee leaders to identify goals, come up with deadlines, and meet on a regular basis. We also tried to lead by example. But how could we tell other leaders how often to meet when they, like us, had many other commitments, both professional and personal?

As cochairs of the GTF, we used the opportunity of the coming of spring semester 2001 to regroup. We took stock of the first semester and concluded that concentrating our forces and working on easily attainable goals was vital not only to the success but to the very existence of the task force.

At this point, the RRR Committee was the GTF success story. It had studied the May term 2000 findings on campus solid waste management and identified and weighed options for a new vendor to pick up and process recyclables. With the help of the GTF standing committee, it had drafted a letter presenting the case to the administration. By early January, the administration had approved the proposed recycling program. The RRR Committee therefore would continue to exist. In addition, we decided to concentrate on the Education and Dining Committees. Rapid progress in promoting campus environmental awareness would help all other GTF efforts, and reducing waste and promoting recycling in the dining halls appeared to be more easily attainable goals than such technically and politically complex issues as cutting energy consumption and eliminating the use of toxic compounds. The other four committees were temporarily abandoned.

Still, we needed something to rekindle enthusiasm and arouse imaginations. Those of us who had attended the PKAL conferences had been inspired by a talk by Nan Jenks-Jay, director of environmental affairs at Middlebury College. Jenks-Jay regularly advises campuses on greening efforts, and we decided to use funds from the ES Rockefeller grant to invite her to meet with the GTF.

Jenks-Jay's visit was a great success. She held brainstorming sessions with each of the three active committees and spoke about her work at Middlebury to the full task force. She also met with the director of physical plant, commiserating with him about the difficulties of his job and sharing information about new technologies and successful practices on other campuses. In addition, she met with the president and helped him to envision the many advantages of a green Illinois Wesleyan. Suddenly the president was asking us if we could be nationally known for our greening like Middlebury. As one member of the GTF said of Jenks-Jay's invigorating talk, "Focusing our efforts on *creating* rather than criticizing plans is a great motivation for me. . . . We want to move forward, not bemoan the mistakes of the past." Later, we used ES grant money to bring other noted environmentalists to campus as part of our Environmental Studies Speaker Series. One speaker was David Orr, whose talk on green architecture had a similar effect as Jenks-Jay in promoting enthusiasm throughout the university for greening efforts.

Developing a Core of Activism

With renewed vigor, the GTF began to coalesce around a shared sense of mission and a new world of possibilities. The RRR Committee began work on implementing the newly approved recycling program. It debated the value, cost, and aesthetics of purchasing recycling bins for offices, dorm rooms, and high-visibility public locations. The Education Committee worked on promoting the expanded recycling program. It developed payroll envelope stuffers, table tents, and large displays explaining the details of the new program to the campus community. Meanwhile, the Dining Committee convinced Sodexho dining services to participate in the campus recycling program and investigated ways of donating unserved food to local homeless shelters, though this last effort ultimately failed.

By mid-spring 2001, word arrived that the new roll-off, the receptacle that would house collected recyclables, would shortly be delivered and that recycling pickup could begin immediately. Although new recycling bins would not be placed on campus until the fall, we had already publicized the program and converted many waste baskets to recycling containers by relabeling them. So as not to lose the momentum, we decided it was time for what we labeled "The Kick Off." We timed the event to coincide with the annual campus Earth Day celebration and the student research conference.

Planning this campus-wide publicity event put the GTF into high gear. In addition to redoubling efforts to label bins and develop educational tools, we discussed other ways of changing the campus culture and identified how important visual symbols would be in this effort. An art major on the GTF designed several greening logos. In one of the most unifying moments of the GTF, the full task force met in a town hall forum to debate the symbolic message conveyed by each symbol under consideration, until all in the room felt proud to have jointly created our "Think Green" logo. The "Think Green" logo was prominently displayed on banners across campus, alongside the university's sesquicentennial banners on the day of the Kick Off. We also invited the campus community and the local media to the Kick Off, where the president and provost proclaimed that the program marked the beginning of a new era at Illinois Wesleyan. Finally, in an act of guerrilla theater, the GTF staged a dramatic symbolic display to encourage waste reduction. On the quad, we heaped one day's worth of garbage for the entire campus. Next to it we placed a livestock feed trough filled with 513 pounds of food waste, the amount generated in the dining hall in just one day.

The publicity we drew to GTF efforts had spin-offs. The student research conference organizers hosted an environmental scientist as the keynote speaker. Eric Pallant, director of the Center for Economic and Environmental Development at Allegheny College, spoke about linking student research to community service through experiential learning courses. The Office of Residential Life made the environment one of the two key themes of the freshman Fall Festival 2001. The Dean of Students' Office subsidized the cost of providing all incoming students with a "Think Green" canvas bag. And the president decided to award an honorary degree to ecologist, environmentalist, and alumna Sandra Steingraber at the fall 2001 President's Day convocation.

While many GTF members were involved in developing and promoting the new recycling program, a core group of activist members had emerged. Although the makeup of this core fluctuated somewhat over the next year and a half, the size remained constant. Almost fifty people turned out for the full task force meetings at the beginning of each semester, and half as many regularly attended committee meetings. However, only a handful of members consistently provided the legwork to turn plans into realities. This group included a few students and administrators, one or two faculty and several staff, in addition to the GTF chairs. Apart from the two cochairs, it was the staff who provided the most consistent and active support. Students and faculty came and went, but the staff who joined the task force in its early days generally remained active throughout the two years of the GTF's existence.

Jenny Hand, a secretary, created virtually all of our educational displays, and almost single-handedly organized training workshops for residence hall advisers and Greek house residents. Jane Randall, another secretary, edited reports and recorded data from dumpster dives organized to measure the success of campus recycling efforts. Chris Kawakita, an admissions counselor, developed a video of our dumpster dive to promote the recycling program. Dave Shiers, the manager of custodial services, and Lawney Gruen, the head of labor crew, implemented the recycling program and constructed the displays of campus waste. Each semester, several staff members helped plan and facilitate the day's-worth-of-garbage, recycling, and food waste displays and the day-long dumpster dives. Without the active commitment of university staff, GTF efforts would likely not have moved off the drawing board. Yet we had never expected staff to play such a central role.

Changing Relationships

It was not simply high-brow academic myopia that had caused us initially to expect limited participation and possibly even resistance from staff. Previous terse interactions with the director of physical plant over energy conservation issues had led us to assume that he was definitely not green. Moreover, a deep divide clearly existed between the academic programs and the day-to-day operations of the university. Although we knew that physical plant personnel in particular felt that students and faculty were treated with undue privilege on our campus, we were yet unaware of the strong hierarchy that informed relationships between the staff and the administration. Nor were we fully aware of the depths of resentment or feelings of insecurity generated by this social structure.

The situation became more apparent as I started to prepare for my May term 2000 "Greening the Campus" course. When I sought help in arranging for guest speakers to share their expertise with my students, the vice president for finance indicated that physical plant personnel were concerned about verbal attacks from "tree-hugging activists." I worked hard to assure the staff that the students would be respectful and equally hard to convey the importance of respect for staff expertise to my students. Still, none of us was prepared for what happened the day that the director of physical plant, Bud Jorgenson, came to class as a guest lecturer.

Jorgenson began by telling the class that the first thing he did when he came to Illinois Wesleyan in 1984 was to develop a comprehensive energy management plan. Over time, he installed variable speed drives in the heating and cooling systems, purchased energy management computers for new and large buildings on campus, replaced old and inefficient steam lines and boilers with more efficient ones and switched much of the lighting from incandescent to fluorescent bulbs. Prior to this, there had been little attention to energy conservation on campus. Still, Jorgenson frequently had to battle with the administration for energy improvements. Most recently, he had argued against putting air-conditioning in the old gymnasium, which was being redesigned as the new student center, because, as he put it, "that building was designed not to have air conditioning. With the proper fans, it could be comfortable." He lost this argument. After silencing the class by mentioning that he had installed a geothermal heating and cooling system in his home, Jorgenson suddenly launched into a vituperative attack on me. In front of a room full of students and his own associate, he angrily recounted his credentials—training in the military, a master's degree in engineering, and thirty years of experience—and decried our efforts to bring in outside consultants to tell him how to operate the campus in an environmentally friendly manner.

It was a profoundly uncomfortable moment, but it was a profound moment of learning too. It forced us to recognize staff members' vast expertise and showed the importance of checking one's own assumptions. It also illustrated the complexity of trying to green our campus. And it drove home how critical it would be to cultivate relationships of mutual respect among the various populations on campus. Simply asserting equality would not be enough.

What none of us could envision at the time was that less than two years later, the director of physical plant would invite the chairs of the GTF to speak about greening the campus at a meeting of regional university physical plant directors. What transpired to turn the tables so radically? How were we able to establish cooperative ties with Jorgenson and other physical plant and staff members?

The key was in acknowledging individual efforts, demonstrating respect, and listening to concerns. We made sure to give credit where credit was due. In their presentation to the administration, students in the May term 2000 class noted the efforts that physical plant staff had undertaken to function in an environmentally responsible manner. The next year, students in a follow-up class designed to study campus energy use reevaluated and endorsed a proposal put forth by the director of physical plant to replace two boilers (see the Illinois Wesleyan University Energy Assessment at <http:sun.iwu.edu/~gtf/energy2001.htm>). The administration had not supported the initial proposal because of the long payback period. The student presentation bolstered Jorgenson's case and offered him moral support. A year and a half later, the boiler project was approved.

We frequently sought the advice of physical plant personnel. As we hammered out the details of how recyclables would be collected, the custodial manager and the manager of labor crew, who served on the RRR Committee, explained to us existing policies and advised us as to what procedures would and would not work. When problems came up, such as a shortage of student workers to handle the recycling, they let us know. We in turn responded to such problems by suggesting ways to ensure a more dependable source of student labor and arguing the physical plant case before the administration.

We sought the expertise of staff members not on the GTF. For example, in trying to identify locations for recycling bins, members of the RRR Committee accompanied custodians through each campus building, asking questions about the content of the garbage and soliciting input on the system we were planning. Months later, when the new recycling program was in place, we invited all custodians and labor crew to a breakfast at which we asked how the system was working, what the problems were, and what could be done to improve the system. Custodians reported that certain faculty and administrators were not recycling, but as staff members, they did not feel comfortable confronting those individuals. We responded by setting up a volunteer system of building recycling coordinators who would help monitor the success of the program and act as the liaisons between the custodial staff and building "residents." Almost all of these volunteers were staff members. Another difficulty that the labor crew reported was that the very success of the recycling program was creating problems. The roll-off was filling so fast that a second roll-off was needed. The GTF responded by raising this issue with the administration, which ultimately approved the purchase of an additional roll-off.

Our efforts affirmed the staff's work, empowered those whose voices were often unheard by the administration, and assured the staff that we were sensitive to their concern about increased workload. Staff members had confided that their resistance to new approaches often resulted from prior experiences with students and faculty who demanded change and subsequently left the staff to shoulder the burden of responsibility for the new project. Our encounters with staff built trust and a shared sense of mission that was often absent between faculty and staff on our campus. They also created a sense of ownership in the success of campus greening efforts among those who, at the end of the day, would be responsible for making sure that recyclables would not simply be tossed in with the garbage. And they helped alert us to some previously unknown issues that we were now able to address.

Yet even as we were establishing good relationships with staff, we encountered many other challenges in trying to green the campus.

Facing the Challenges

Chief among the problems we faced during the two years of GTF work were a general resistance to change, a lack of financial commitment for greening efforts, a shortage of leadership within the GTF, and, as a result of all of these, periodic feelings of burnout among core members.

Resistance to Change

Resistance to change was manifested in a number of ways. First was the issue of aesthetics. We had difficulty overcoming the university's opposition to locating recycling bins in prominent places. This remained the case in spite of the fact that convenience is often cited as the most important factor affecting the success of recycling efforts and that the GTF maintained that prominently displayed recycling bins were necessary to convey symbolically the campus's commitment to environmental stewardship. In our discussions with various administrators, we insisted that prospective students were less likely to view the presence of recycling bins as an eyesore than they were the absence of recycling bins. Yet on the very day of the Kick Off, after many congratulatory remarks from the president, the provost, and others on the GTF's success, the vice president for finance privately informed us that we would have to reconsider the issue of buying recycling bins for the quad. The university had spent millions of dollars to make its buildings attractive and did not want anything to detract from their appearance. Although we had promised to pay the full costs of the recycling receptacles (approximately $20,000) with ES grant money and had carefully selected receptacles to coordinate with existing campus waste baskets, we were unable to win over the vice president for finance until we solicited help from the assistant to the president. Ultimately, our persistent efforts changed the mind-set of the university administration. Built-in recycling bins were included in the new library and student center completed in 2002.

Resistance to change also posed problems in convincing certain staff to change their behavior. For example, the director of publications, printing and mailing services, who was knowledgeable about the toxins in ink, thought that recycling paper in an environmentally friendly manner was not possible, so she had refused to recycle any inked paper. She also believed that it was not financially possible to commit the university to purchasing any quantity of 100 percent postconsumer, bleach-free paper. As momentum built on campus for the GTF's efforts, however, she began to change her attitude. She began to attend RRR meetings and helped to investigate options for purchasing recycled paper.

In brief, though some individuals on campus did not care about environmental sustainability, others did not necessarily have anti-environmental views; they were simply uncomfortable with change.

Communication and respect were the key ingredients for success. Rather than demanding cooperation, we learned that it worked better to ask for assistance, offer help, and educate people about the environmental benefits of change.

Lack of Commitment

Another difficulty we faced was the lack of commitment by the university to routinely finance and institutionalize campus greening efforts. Although the university had invested money to improve aspects of campus energy efficiency over a decade ago, such improvements were made largely for financial rather than environmental reasons. When the GTF was established, no money was allocated to it. It was clear that the administration would be more likely to support GTF proposals if funds were not a concern. As a result, we used external ES grant money to purchase the initial roll-off, recycling bins, banners, and other items. By the second year, members of the task force had become frustrated with the university's lack of tangible commitment to long-term support of campus greening efforts. The ES program had already contributed close to $30,000 for the expanded recycling program, and the grant term was drawing to a close. Frustration over funding was particularly demoralizing for those who had been most active on the GTF. Some felt it was time to publicize the university's failure to finance the greening effort to local media and the student body. As directors of the GTF, however, we had consistently tried to maintain a nonconfrontational approach, believing that aggressive approaches generally backfired. We decided to raise the issue directly. In a private letter to the administration in which we requested funding for a second roll-off and additional recycling bins, we pointed out precisely how much ES money had been used for GTF purposes. The administration's agreement to pay for these additional needs was an uplifting victory for the core members of the GTF. On the larger issue of long-term funding to institutionalize the campus greening movement, however, the university has yet to make a commitment.

Lack of Leadership and Initiative

As cochairs of the GTF, we have shared the frustrations of resistance to change and shortage of funds. Equally frustrating to us has been the lack of leadership and initiative within the task force, and particularly among faculty and students. When the GTF began, we had hopes that faculty members and students would take leadership roles. We were disappointed early on when we were unable to recruit the powerful, progressive faculty voices we had expected and when only three of thirteen ES faculty joined the GTF. Most faculty members gave only moral support. Even the few who volunteered to lead a committee found themselves unable or unwilling to devote time. Their committees languished; active members who had raised excellent suggestions and potentially active members drifted away for lack of leadership. This problem persisted even after we reduced the number of committees and consolidated forces. Student initiative too was much more limited than expected. Even the most active staff members on the GTF would not—or felt they could not—assume leadership positions. By the second year, it was clear that successful efforts depended heavily on the leadership of the GTF cochairs.

Added to our frustration over limited faculty and student participation was the fact that with success came vastly increased demands. For example, as GTF cochairs, we received frequent telephone calls and e-mails: requests for more bins, inquiries about the recycling program, and numerous reports of problems. Fraternities and sororities had not been included in the expanded recycling program because they were not on the labor crew's solid waste pickup route, yet Greeks wanted to participate. The flood of recyclables overwhelmed the labor crew. In desperation, the labor crew at one point dumped the recycling in the trash—and an outraged student reported to us that Illinois Wesleyan had stopped its recycling program! In addition, we handled numerous issues related to publicity. For example, the school newspaper ran several articles on the GTF's work, but reporters were not always politic about the way they presented issues and did not always know about ongoing negotiations. As directors of the GTF, we found ourselves in the position of trying to stamp out numerous smoldering fires before they turned into infernos. We felt further challenge by requests for action on greening issues other than recycling: calls to reduce campus junk mail, replace bathroom paper towels with electric dryers, and eliminate campus use of chemical pesticides and fertilizers—which we were unable to address.

Even when we were not putting out fires, we were overwhelmingly busy in the massive undertaking of fully implementing the recycling program. Over time, our efforts allowed us to create a near complete infrastructure for recycling, perfect a monitoring system able to identify problems with the recycling system in specific buildings, and produce educational programs appropriately targeted at different sectors of campus. The whole process, however, was extremely labor intensive.

In brief, we concluded we were unable to do it all. We had only begun to address a few of the many issues of sustainability we had initially identified in August 2000. At the start of the second year of the task force, we drafted a detailed proposal requesting that the university hire a full-time coordinator of environmental affairs. We shared this proposal with the top administrators most sympathetic to our efforts. However, due to the overall financial climate, the proposal was rejected.

Burnout

For those of us most active on the GTF and most committed to environmental improvement, the general resistance to change on campus, the lack of sustained commitment for greening efforts on the part of the administration, and the shortage of leadership and initiative within the GTF led to periodic feelings of burnout. As cochairs of the GTF, these feelings were sometimes intense. We were also directing the new ES program in addition to teaching six classes a year. One of us was not yet tenured, was still developing courses, and had the added responsibilities of servicing two interdisciplinary programs and caring for a young family. The other had an extremely heavy advising load and responsibilities to students pursuing admission to graduate and medical school, as well as family commitments.

To keep us going, we learned that we needed to delegate responsibility, reduce expectations, and scale back our immediate goals. We replenished our depleted enthusiasm by attending ES conferences, such as the Ball State Greening the Campus Conference, where we made contacts with like-minded individuals and were inspired by the stories of how others persevered and overcame obstacles. We garnered strength from each other and spelled one another when one of us was feeling particularly overwhelmed. We also tried to integrate our teaching responsibilities with our efforts to green the campus.

To make up for the absence of voluntary student initiative on the task force, we offered the second experiential learning course, which focused on analyzing campus energy consumption, though with mixed results. The course allowed us to use student research skills to investigate an important aspect of campus life to which the GTF had been unable to devote attention. Students conducted a detailed campus energy audit, produced a briefing book with policy proposals, and developed educational posters and pamphlets to promote campus energy conservation. Nevertheless, they were frustrated by the intensity of the workload, and though most were ES minors, they, like many of their counterparts on the faculty, lacked the passion for the greening effort. In the end, our disappointment with the students' attitude and the challenge of teaching this course further added to our own feelings of burnout.

Other active members on the GTF also experienced burnout. Lawney Gruen bore the brunt of hauling the recycling to the roll-off, day-in and day-out, regardless of how short his crew was on a particular day. We sensed his frustration and tried to support him by frequently inquiring about his work and being advocates for his crew.

Jenny Hand, the muscle behind our environmental education programs, nearly quit the task force in frustration after its first year. She felt she had become a "one-man show." Several students would agree to help out but would not follow through, and other GTF members would not even offer assistance. In addition, university policy required her to make up work time spent at GTF meetings or use vacation time to attend dumpster dives. We averted this potentially major loss by discussing the problem with Hand and coaxing others to take more responsibility for educational work.

One of the few students who heeded our call for greater initiative ultimately felt demoralized. Becky Heine volunteered to organize the spring 2002 dumpster dive. She worked hard to coordinate the collection and labeling of recycling from each building, publicize the event widely, and recruit students. But on the day of the dumpster dive, she was crestfallen: only a few new students participated, and most of the job was performed by the GTF's faithful few. However, she garnered strength from the enthusiasm, determination, and commitment of that same faithful crew—as did we all, periodically.

Moving toward Sustainability

In April 2002, the GTF issued a final public report, recording the university's environmental achievements and reiterating the case for a coordinator of environmental affairs. It also recommended the establishment of a permanent Committee for a Sustainable Campus (CSC). The latter recommendation was accepted; the former has not been.

Nevertheless, we are guardedly optimistic about the next stage of our campus greening effort. We have developed a core of individuals committed to greening Illinois Wesleyan's campus and have established strong relations with physical plant and other staff. The campus community and, notably, the administration are beginning to become keenly aware of the environmental implications of its daily actions. In addition, in order to cope with a shortage of leadership and maintain momentum, we have decided to concentrate our efforts and focus on a couple of carefully defined, reasonably achievable goals. (In contrast to the GTF, the CSC will have no subcommittees and half as many members, and will continue the work that the GTF started, before pursuing new goals.) However, the long-term prospects of any effort to transform IWU into an ecologically sustainable campus will depend on whether the university chooses to hire a full-time coordinator of environmental affairs. In the meantime, the speed and extent of progress may depend in part on the speed with which the ES program develops. When IWU is able to offer an ES major, we hope to attract environmentally activist students to our campus, eager to bring about change.

Conclusions

Our experience suggests that while small liberal arts colleges without a history of environmental concern face serious challenges to promoting environmental stewardship, successful institutional efforts are possible. It is important to build and maintain momentum at critical times, view the process as an ongoing endeavor requiring well-paced and consistent effort, make use of ES resources, and stay in touch with like-minded individuals at other universities. Perhaps most significant, it is important to develop relationships of understanding, respect, and mutual support among various sectors of the campus community, paying particular attention to university staff. Over the long term, proactive administrative support is essential for the university to be genuinely green.

As we have moved into the implementation stage of recycling, this last factor has become clear. Although the physical infrastructure is now mostly in place and many faculty, staff, and students are aware of the recycling program, our most recent assessment of the program, in November 2002, indicates that the university still has a long way to go to meet its unstated goal of a 50 percent recycling rate. Without continual university-wide educational efforts and consistent commitment from the administration, substantial energy by the newly created Committee for a Sustainable Campus will be necessary to maintain and improve on our modest achievements.

3

No Longer Waiting for Someone Else to Do It: A Tale of Reluctant Leadership

Peggy F. Barlett

Emory University is a private, medium-sized research university of 6,000 graduate and professional students and 5,000 undergraduates in the liberal arts. The 630-acre campus is fifteen minutes from downtown Atlanta (whose sprawling metropolitan area numbers around 4 million), in a mixed neighborhood of historic homes, suburban shopping malls, and some dense urban corridors.

Frustrated, I waited over three years for someone else to step forward to galvanize campus action toward sustainability at Emory University. Though valuable efforts were underway—in recycling, in reducing the harm of new construction—no one was fostering a more profound campus-wide questioning, based on the awareness that our daily lives contribute to the degradation of the earth's natural systems. The campus ethos seemed untouched by the front-page news of Atlanta's declining air quality, water pollution, traffic congestion, and deforestation. In the summer of 1999, a decision to build a disputed road through beloved campus woods generated anger that simmered for months among faculty who normally expressed no environmental concern. For me as an anthropologist, the question of how my workplace might change was connected to a larger question about how the transformation of Western industrial society toward sustainability could come about. How do we step forward to so radically different a future?

Margaret Mead taught us that cultural change is led by small groups of thoughtful people, working together. Small groups at Emory were already working on university committees, and an Environmental Studies Department was just forming, yet broad opposition to the road had no mechanism to come together, no way to reflect on the difficult tradeoffs involved in the decision, and no way to channel our love of the woods and concern for the regional environment more constructively.

Reluctantly, I decided to step forward to see if it were possible to nurture the formation of small groups of thoughtful people to work toward campus change. This chapter is an abbreviated account of three years of work, and I include my doubts and disquiets as well as my delight, in the hope that others who hesitate will find the encouragement to step forward into their own unwelcome spotlight.[1]

If change comes from small groups, then how we foster small groups matters. As faculty, administrators, and staff, few of us think about such things, nor do we act in accord with the philosophy that change comes through relationships. A university such as Emory is really a "small city," with complex connections and disconnections among graduate, undergraduate, and professional schools. With no faculty union or cohesive tradition, the institutions to undertake such cultural change were not in place. I could only dimly perceive the need for some preliminary organizational steps to build trust, share information, and find visible projects to raise environmental awareness.

In this chapter, I recount how several new organizations emerged organically over time at Emory and describe the steps we took to foster effective group process. I begin with the creation of the Ad Hoc Committee on Environmental Stewardship and its two major projects: efforts to restore a small campus woodland, adjacent to the quadrangle, and a campus-wide environmental mission statement. The Ad Hoc Committee is a broad coalition of faculty, students, staff, and alumni, and its efforts were directed toward the larger campus ethos and operations. Other projects followed, directed more at faculty and the teaching and research dimensions of sustainability: the Faculty Green Lunch Group and the Piedmont Project for curriculum development. Though there have been bumps in the road in each of these activities and their continuation is not guaranteed, with hindsight I can say that the rewards have been enormous, the personal growth substantial, and the responses of others both inspiring and gratifying.

The Ad Hoc Committee on Environmental Stewardship

I first heard about Indiana University's Council on Environmental Stewardship at the 1999 Orion Society Conference, "Fire and Grit," an inspirational summer gathering of nature writers and grassroots groups from around the country. The language of stewardship resonated for me. It was an important concept in my childhood religious upbringing, and it seemed to me that it would provide legitimacy in the Emory context, where campus activism is rare. Emory is a Methodist institution with a well-known theology school, and no one can argue with the assertion that we are stewards of valuable resources—not only our monetary endowment, but also lovely forests, several creeks, and the gentle hills over which campus buildings are clustered. The Orion Conference also introduced me to the Penn State Campus Environmental Indicators Web site (see chapter 1; <http://www.bio.psu.edu/greendestiny>). Once I saw its detailed discussions for improving campus operations and clear recommendations for first steps and later steps, I realized that the rationale, scientific knowledge, and practical information there would let us begin at Emory. I thought, "With this to fall back on, we *can* move ahead."

The controversy over the road created the urgency to establish an organization to facilitate information sharing and action. I began by floating the Indiana University Council on Environmental Stewardship idea with about a dozen colleagues and friends. They were positive but urged that the group keep a low profile, using the label "ad hoc committee" in order to seem less threatening. Heartened by a sense that such a group might be useful, I sent out an e-mail invitation to all the faculty, staff, and students whom my friends and I thought might be interested (about seventy people) and asked each to pass the word on. My first hurdle was the decision about who should sign the e-mail. None of my friends was interested in helping to organize the group, so I decided it was more honest to sign it alone. My hope was that a group would emerge to share leadership and my name would fade into the background.

On a late September afternoon in 1999, twenty-one people gathered to explore the possibility of an ad hoc committee. My second hurdle was how to facilitate the meeting. If the group was diverse, I was afraid people might be wary (this concern turned out to be valid), and I was anxious that students and staff not feel dominated by faculty, who love to talk. I wanted participation to be broad in order to release the creativity of the group. With considerable trepidation, I decided to take a strong facilitator role and to use ice-breaker techniques that I had learned in group dynamics workshops.

As the group gathered, I passed out scrap paper and asked them to jot down responses to two questions: "What concerns do you have today about environmental issues at Emory?" and "What is the vision you'd like to see for the future—what are some pieces of how you would like it to be?" When it was time to start, I shared a brief introduction about my sense of the ferment on campus and the need to educate ourselves. I explained the Indiana University model and wondered whether an organization to promote environmental stewardship was right for Emory at this time. I emphasized that environmental engagement need not be a Puritan hair shirt, but it might be an opportunity for us to move in some satisfying directions. Then I posed two questions for the rest of the meeting: Who are we, and what do we want to do? I suggested that we go around the room with introductions and in addition to sharing names and university affiliations, that we share something from our list of concerns. This whole introduction took about five minutes, then shifted the focus to the rest of the group. Some people spoke calmly about air pollution or population growth, but others shared with deeper personal language about loss of biodiversity or how the university's use of resources was personally painful.

In order to break up the somewhat stiff interaction in the room, I then asked that each person stand up and find someone in the room he or she did not know and requested each pair to spend five minutes introducing themselves a bit further and sharing some part of the second question about their visions for Emory's future. To my relief, people accepted this unusual exercise, and the room babbled with voices. I was pleased to see a stiff faculty member conversing comfortably with an undergraduate student whom I believed knew very few people in the room. When we reconvened to brainstorm about next steps, the discussion was lively, most people contributed, and I think these two exercises helped to build greater comfort within the group. One volunteer offered to create a listserve for future communication (which worked very well), and another urged that we meet again to get to know each other better.

At the second meeting three weeks later, we skipped the second group-building exercise, but we did introduce ourselves again in a lengthier manner. Attendance was about the same, though with some new faces; more undergraduate and graduate students and fewer faculty came, and several new staff members attended. We also rearranged the furniture to put chairs openly in a circle rather than sit behind a U-shaped table, which seemed to make interaction more relaxed.

Looking back, I realize that our second meeting was an important lesson to trust the wisdom of the group. I had been reading business and organizational development literature that emphasized systems thinking and learning organizations, but I had considerable skepticism about the stories I had read (Capra 1996, Jaworski 1996, Katzenbach and Smith 1994, Senge 1990, Wheatley 1992). That day, the discussion ebbed and flowed for over an hour, exploring what we might do as a group and on which dimensions of Emory's functioning we might focus first. Time was running out, people would soon start to leave, and no consensus was emerging. I was concerned that attendance for a third planning meeting would be low. A landscape architect from Facilities Management was talking about some efforts on campus and was using official names for streets and locations—names that are not in common use. Faces were blank, so I stopped the flow of the conversation to check one of the terms: "How many of us know where Baker Woods is?" Only three people raised their hands. The speaker shifted gears a bit, tension eased, and then a graduate student from public health suggested that we get someone to give us a tour of some of these forests. "How can we think about good stewardship of our resources, when we don't even know what they are?" she asked. Bingo. We'd found a clear next step. The group loved the idea. I knew of a prestigious biology professor, recently retired, who would be perfect to lead us, and the energy in the group rose. We little imagined that such a simple suggestion would lead to powerful and long-lasting results.

At the time, we also planned several other foci of action, but they were never realized. I also worried about our failure to coalesce around one coherent and visible project. In addition, John Wegner, an ecologist in environmental studies, had spent part of most days in the summer of 1999 watching and guiding the removal of trees and the construction of the road, but few knew of his ameliorative efforts. I think now that there would have been benefits had the group found a single, clear focus, but our diversity encouraged a broader range of activities to emerge later. John's decision to engage intensively with facilities management and construction personnel, however, built trust and later support for environmental issues. I also was disappointed that a team of senior faculty colleagues did not emerge to join me in those early meetings. I thought about giving up; though exploring the woods sounded like fun, I was not sure that it would head us in a useful direction. A friend whose own time was already overcommitted gave me encouragement: "Just keep calling the meetings, and a year from now things will be different." Dubious, I filed that advice away.

Reluctant Visibility

There were lots of reasons I felt I was not the appropriate person to lead campus environmental change. As a social scientist, I knew only a little about ecology, the history of the environmental movement, and current issues such as global climate change or acid rain. My anthropological teaching about Latin America led me to feel reasonably competent about issues such as deforestation in the Amazon, but I did not know much about light bulbs or even what is a VOC (volatile organic compound). When engaging other academics in debates about carbon trading or genetically modified foods, I found it hard to be persuasive. Even in the realm of behavioral change, presumably closer to my social science training, there were worlds of applied psychology, organizational development, and persuasive homiletics that might be brought to the service of environmental causes but were closed books to me. Surely, I felt, other scholars whose work was more centrally related to environmental issues would stand up and lead us, and I could play an energetic but supportive role.

My twenty-three years of experience with Emory politics and governance, in committees and as department chair, also made me hesitant to step outside my disciplinary expertise. Emory has a tradition of relatively weak faculty governance, and our committee system is cumbersome and often ineffective. Open faculty meetings have not been locales for thoughtful dialogue, and many good ideas have failed to gain support. The voices of a curmudgeonly, cynical minority are loud, and standing up for positive movement—even a campaign of action—that might involve the daily lives of peers is virtually unknown. In addition, women's voices are less often heard in public discourse at Emory (though in a series of dramatic changes, that pattern has shifted in recent years). I felt that a woman leader—and a liberal social scientist to boot—would be less effective in the relatively conservative Emory context.

At the same time, my intellectual interest in agrarian economic development was intrigued by the emerging international paradigm of sustainable development and noted the ways it harmonized with fundamental anthropological understandings. Active in building a neighborhood watershed alliance, I was also stimulated by learning more about the city and bringing social science insights to bear on urban environmental dilemmas. As the fall semester rolled on, and other, more likely leaders were too busy or unwilling to step forward, I remained convinced that the time was right for the Ad Hoc Committee to contribute to environmental movement on campus. I listened to another friend who argued that campus action needs a point person, and that person should be me: "No one else can do it right now." With the privilege of tenure and strong networks to various parts of the university, I decided that if someone like me was unable to set aside for a time the mandates of publish or perish, who could? Slowly, I built up the courage to step out front and began to articulate more publicly the vision for Emory and to accept a more visible role in campus publications and in dialogue with decision makers.

Lessons from Grounding Ourselves in Place

On a misty Saturday morning in November 1999, with a vee of geese honking overhead, a dozen individuals drawn from almost every professional school and division of the university met for our first woods tour. The walk was magical, the learning about the place where we work was rich, and the experience out-of-doors was the kind of time-out that builds camaraderie. It also greatly deepened our appreciation for the Baker Woodlands, a three-and-a-half-acre patch of woods we toured (which I had personally referred to in the past as "the gulch").[2] With over 100 plant species, it is a lovely, relatively healthy piece of Piedmont forest. But it faced serious invasion from English ivy and streambed erosion from new pavement upstream (a new parking deck and an addition to the library were the main culprits). The tour leader, Bill Murdy, expressed his dream of major ivy removal to protect the rare wildflowers now slowly disappearing.

There was some enthusiasm for the idea of an ivy pull, and the whole Ad Hoc Committee embraced it. New people came forward to help with publicity and preparations, and in February 2000, eighty people gathered on a Saturday morning to load truck after truck with ivy. Several more such restoration events, together with planting of native azaleas and other shrubs, led to a glorious Baker Woodland in spring 2002, in which many new blooms of trillium, oxalis, sweet shrub, and wild azalea amazed us all.

Regularly scheduled woods walks became an important way to have fun while learning important environmental lessons. Each time, new people joined the group, drawing folks from all over the campus. But the tour leaders' time was limited, and I mused about ways to reach larger numbers. Someone suggested a self-guided walking tour, and so I asked eight experts from around the campus to write a small pamphlet. All of them gave up precious August days to work on it (Barlett 2002). With the support of the Office of the President, we published a brochure that outlined a walking tour of ten campus sites of particular environmental importance (both challenges and successes). The brochure is available on-line: <www.environment.emory.edu>.

Our experience supports the work of Davey, Earl, and Clift (1999), who found that learning about local impacts of environmental processes is the best way to engage university stakeholders. Several classes and new student orientation activities now use the walking tour, and many participants report that it profoundly shifts their awareness of the campus world through which they walk each day. During one tour with administrators, we stood in a parking lot and learned the way creek organisms are harmed by the heated water of summer rains. The person standing beside me exclaimed, "Wow, I never thought about that." Her eyes widened, "But, of course, it would do that." A pause: "This tour is so *important*. We need to find a way to bring this to the board of trustees." The brochure is now in its second printing and has become an important tool for environmental awareness on campus.[3]

The Wisdom of the Group, Part II

Building on the camaraderie of the woods experience, the Ad Hoc Committee began to take stock of university environmental functioning, using the Sustainability Assessment Questionnaire (SAQ) developed by University Leaders for a Sustainable Future (Calder, Clugston, and Rogers 1999) <www.ulsf.org>. To strengthen the dialogue in our first SAQ discussion, I felt we needed a good range of faculty, staff, and students to be present. I made several dozen personal calls to people I thought would have valuable information and whose participation in the conversation might energize planning. About thirty people showed up, and it was the first time that many of us became aware of the substantial efforts already underway in several university units. Alternative transportation had programs to support free bus passes for employees, carpools, van pools, electric vehicles, and natural-gas-powered shuttle buses. Recycling was gaining administrative support. An important baseline study of Emory's forest resources had been completed years ago, and its continued effectiveness in guiding campus planning was discussed. A limited range of appropriate courses was found in various schools.

No consensus emerged, however, for next steps for the Ad Hoc Committee. Toward the end of the meeting, a student spoke up in frustration, "We can't really assess how well Emory's doing because we don't really know where we want it to go." Others supported this idea: "Yes, we need a policy!" A subcommittee was born to study other schools and return with a draft environmental policy. Though the minutes for February show that many other efforts were underway (plans to attend a regional Second Nature workshop, an Earth Day vendors' fair for office managers, an art show), the decision to move toward a policy, later renamed the mission statement, had far-reaching impacts on campus awareness.[4]

Building grassroots awareness was an important next step for the Ad Hoc Committee, and the mission statement gave us the opportunity to engage with more constituencies, strengthening the process of cultural change, and discovering more about how different sectors of the university see environmental issues. Our hope was that even if the formal adoption of the mission statement floundered from political opposition for some reason, the *process* of consultation would raise awareness, itself a useful step (Mumford 2001).

To begin drafting our mission statement, the guiding subcommittee of eight (four faculty, two staff, and two students) used the International Institute for Sustainable Development's Policy Bank to study examples from other schools <http://iisd1.iisd.ca/educate/policybank.asp> (Mumford 2001). We consulted with the whole Ad Hoc Committee on general length and tone, wrote a draft text, and then revised it with input from the larger group. With the revised draft in hand, we solicited formal support by e-mail for the mission statement process from a wide range of Ad Hoc Committee supporters, senior administrators, faculty, and student leaders. As the names first trickled, then poured in, we copied a list of thirty-seven supporters on the back of the draft text. We learned quickly that in our consultations, people scanned the names right away. It was reassuring to them that a university senate president, distinguished faculty in law, public health, medicine, and theology, as well as several key facilities management leaders and graduate and undergraduate students were willing to be publicly supportive.

The Mission Statement Consultation Process

In retrospect, our decision to use a consultation meeting format to build support for the campus-wide mission statement turned out to be very important, though we also made some mistakes. Over eight months, the consultation process involved twenty-two formal meetings with groups from all parts of the campus: food service, campus life, libraries, human resources, business management, facilities management, purchasing, plus the campus-wide Employee Council and Student Government Association. Meetings at Emory's two-year affiliate, Oxford College, were held separately with faculty, staff, and students. Support was requested from all relevant University Senate committees. In general, the Ad Hoc Committee asked for ten to thirty minutes in an already-scheduled meeting of the unit, usually with its leaders or management staff. We did not find an easy way to meet with the rank-and-file of most units.

One place I was comfortable taking the lead was in using my longevity at the university, my status as a senior faculty member, and my professional ties with people all over campus to gain access to busy meeting calendars. I decided to meet privately with all the deans because I knew a number of them personally, and I solicited their advice about how best to consult with their faculty and promote environmental action within their schools. In retrospect, for those deans and administrators who did not know me personally and were less aware of environmental issues, it would have had more impact to meet with a heavy-weight group of Ad

Hoc supporters rather than one person. The deans advised against discussing the mission statement draft in a regular, full faculty meeting, recommending instead an open invitation to a lunch or breakfast gathering over a previously circulated document. We followed this advice; in most cases, a faculty member within the school convened the gathering, and anywhere from half a dozen to twenty faculty in each school discussed the draft text. Suggestions for wording changes led to some valuable revisions—and more names of supporters.

The consultation process itself, developed from the ideas of Karen Mumford in Environmental Studies and Mary Elizabeth Moore in Theology, involved five steps, and their sequence turned out to be critical to the surprisingly positive response. First, the two to four Ad Hoc Committee members presented a brief preamble, introducing the history of the committee and how the mission statement came to be written. We meanwhile passed around a sign-in sheet asking for names and e-mail addresses. Then introductions were requested to legitimize participation from all present. In addition to saying name and area of responsibility, each speaker was invited to "name an environmental issue that concerns you personally, maybe something here in Atlanta or at Emory or something international." The person responsible for the preamble would then model what we were looking for: "I'll start. I'm Karen Mumford from the Environmental Studies Department, and an issue that concerns me today is the loss of trees here in Atlanta as we continue to grow so fast." We found that virtually everyone had something heartfelt to say about environmental concerns. By the time introductions were finished, we had no need to make a case for why Emory should adopt a mission statement: the case was made for us!

Our third step was to ask how the unit had already responded to environmental concerns. To our surprise, most units were proud of several actions and were delighted to tell us about them. We too were gratified to learn of these activities, and it shifted our sense that "Emory hasn't done much" to "We've done more than we thought." Karen wrote up all these activities, and we tried to get campus publications to do stories on them, but with only a little success.

We found it important next to read the whole mission statement aloud, because its rhetoric was appealing, and a review of its points helped focus discussion. Time for comments was short or long, depending on the meeting's agenda, and then we closed by distributing a copy of the "Tufts Dining Strategic Plan" from *Greening the Ivory Tower* (Creighton 1998, 292–299). This handout provided a clear example of how the food service at Tufts University took a general mission statement and translated it to specific outcomes, strategies, and action steps. Several administrators found it helpful to have a concrete example of where a mission statement might lead. At an appropriate point during the meeting, we asked for support and invited new signatories. To make that process even easier, we sent a follow-up e-mail to all who attended the meeting, thanking them for their time, asking for any further thoughts, and offering information on how to subscribe to the Ad Hoc Committee's listserve. We got only a few new participants in this way, but it seemed to help spread awareness of the group and its activities.

Some of the consultations were friendly conversations, and people seemed curious and open to the Ad Hoc Committee's presentation. Other groups were defensive or wary. Our open style and the fact that the consultation started with lots of listening on our part usually shifted the atmosphere. Some groups were hurried and distracted, but discovering that colleagues were worried about air or water quality—and maybe even cared passionately, suffering along with an asthmatic child or a vulnerable elder—encouraged the group to pay more attention. Most consultations created an atmosphere in which participants recognized that the issues are important.

One of the surprises for me in the consultation process was that our work sometimes brought relief. Several people came up to us afterward and expressed gratitude that "someone is finally *doing* something" or "this is so overdue; thank you for bringing up these issues." Many workers at Emory have in fact heard worrisome environmental news and want to act. Learning of the existence of the Ad Hoc Committee was a relief to the worry or guilt they feel, and we tried to follow up with these individuals to offer opportunities for them to act. It was affirming to me to learn there were others in unexpected quarters of the university who were waiting, too, for someone to take the lead.

Spring semester 2001 saw the completion of the consultations, and the mission statement was placed on the university senate's agenda for a formal vote of adoption in February. One mistake we made was to streamline the presentation to the senate in deference to its crowded agenda. I was delegated to be the sole spokesperson, and we thought our careful grassroots work and support by two senate committees and by the senate president (and the informal signal of support from both the president of the university and the provost) meant the vote would be easy. Our list of supporters now had ninety names drawn from all parts of the university, including endowed chair professors and a vice president. Unexpectedly, a representative from the medical school, a unit of the university that had shown little interest previously in the mission statement and had declined two requests for meetings or consultations, read a lengthy statement in opposition. My naming of all the groups that supported the mission statement was outweighed by this strong counter-voice. The medical school argued that affirming environmental priorities might inhibit the rapid physical plant growth they felt necessary for their future academic excellence. This threat was not balanced by any particular awareness of the links between health and the environment or benefits to the medical school of embracing a greener approach to hospital operations or even medical school curriculum. It was also clear that our decision to make a quick presentation left some senate members without any concrete ideas of what kinds of environmental change might ensue from adopting the document. After some difficult discussion, the vote was tabled until the March meeting.

Several Ad Hoc Committee members, and especially those with ties to the medical school, then began a series of hurried visits with key department heads, and there were intense negotiations over the wording of the document. In retrospect, we should have tried harder before the senate vote to find medical school leaders for formal consultation or informal dialogue. The March meetings produced several wording compromises, the new document was reluctantly approved by e-mail by the Ad Hoc Committee supporters, and a new presentation was prepared for the senate. This time, the mission statement effort was described by four individuals, and the truly broad nature of the supporting coalition was more evident. Examples from green computing, solar power options, and the State University of New York at Buffalo's green office forums illustrated positive examples of change. The vote was nearly unanimous in favor, and the senate went on to ask the president to appoint an implementation task force, to recommend a management system to turn our fine rhetoric into reality. Exhausted, we ended the spring semester with a late-afternoon celebration and a sense of real progress from two years of Ad Hoc Committee effort.

Follow the Energy: The Faculty Green Lunch Group

While the woods walks, ivy pulls, and the mission statement effort were emerging as useful avenues to build campus awareness, I was troubled that only limited numbers of faculty and students were involved. On many campuses, students provide the real energy toward campus greening, but student environmental interest was not strong at Emory in those years. Faculty mentorship is critical to supporting student action, which suggested the need to foster faculty involvement. Though perhaps some eighty faculty were quietly or openly supportive of the efforts of the Ad Hoc Committee, most felt their scarce time could best be used for research and teaching.

Discouraged, I discussed this dilemma with Howard Frumkin in the school of Public Health, and he suggested we "go where the energy is" and form a "faculty environmental interest group." Lance Gunderson, the head of the environmental studies department, pledged his support and the cost of box lunches for two meetings. The provost kicked in an equal amount of money, which allowed us to set up four dates for the spring 2000 semester. The Faculty Green Lunch Group was born (Barlett and Eisen 2002). The lunchtime format spanned an hour and a half, bridging two teaching periods. The format evolved into a twenty-minute presentation by a faculty speaker about current research or teaching related to environmental issues, followed by discussion. Attendance was most commonly between fifteen and twenty, though once as high as twenty-nine. Slowly, a collegial group solidified, leading to broader efforts to affect teaching and research than I would ever have dreamed.

Creating community requires that we know and trust each other. With such a diverse group of faculty, I pushed a tradition of introductions with queries. Though some faculty have gently suggested we can dispense with "the queries," I have just as gently encouraged us to start each meeting by saying our names and departments and answering an open-ended question that allows for self-reflection, creativity, or humor. For example: "What was something you feel grateful for today?" *Getting my lawn mowed after three weeks!* "How did you engage with the natural world over the break?" *I finally got to see the 400-year-old poplar in North Carolina.* "What was something interesting you learned recently?" *That 17 percent of undergraduates think "a lot" is one word* (Barlett and Eisen 2002, 5). Putting all voices out into the room allows shyer people to contribute equally and lets us get to know one another without the competitive posturings that can sometimes afflict faculty discussion groups. The way faculty chose to introduce themselves also acknowledged the whole person, with family travails and outdoor experiences, as well as intellectual interests.

The dynamism of the discussion and the loyalty of the following, even among those who have teaching or committee conflicts and cannot come regularly, was unexpected. "It's really a community," said one. Faculty were interested in being educated on issues, especially by peers who are willing to talk across disciplinary boundaries. Once a semester, the topic focused on a teaching dilemma, and these discussions tended to have the highest turnout. We also had thirty people show up for a post-lunch tour of one of Emory's new green buildings.

The Piedmont Project at Emory

The Faculty Green Lunch Group became the seedbed for the Piedmont Project, Emory's effort to green the curriculum. I was mulling over the question of how to foster deeper engagement with environmental issues in the curriculum when I went out to Arizona to participate in the Ponderosa Project workshop in May 2000 (see chapter 3). The possibility of a course requirement for all students was nil at Emory. We had just completed a painful curriculum revision, and inserting a new requirement would not happen soon. I liked the way Northern Arizona University (NAU) wove sustainability issues into the fabric of intellectual life. Would my colleagues at Emory be willing to engage in two days of lectures, discussions, and pedagogical exercises? Would enough people be interested in changing their courses?

Arri Eisen, a faculty member in the Biology Department and head of the Science and Society Program, joined me to draft a proposal for a summer program for 2001, to support the development of twenty new courses (or course modules), and we shared it with the Green Lunch Group. Nearly a dozen people expressed immediate enthusiasm. The University Teaching Fund supported the proposal, and what came to be known as the Piedmont Project was born. In the three years we have run the project so far, we have followed closely the NAU model, and Geoffrey Chase and Paul Rowland from NAU facilitated the first workshop. The two-day workshop and kick-off dinner are held immediately after graduation. Three or four resource experts describe how environmental issues connect to their fields, and many small breakout group discussions allow the twenty participants to get to know each other, broaden their thinking about both content and teaching methods, and reflect together about what are our ideal educational outcomes.

A very enjoyable part of the workshop are the woods walks each day after lunch, led by Eloise Carter, an enthusiastic ecologist of the Piedmont. The workshop is held at the edge of campus, with nearby woods to showcase local flora, the damage of invasive species, and water pollution issues. Once again, our deeper connections with the *place* where we live and the joy of spending time outdoors not only strengthens environmental knowledge but deepens the connection among group members.

A follow-up meeting in August involved a field trip to Oxford College and some fascinating aquatic ecology in the campus pond. "It was an intellectual feast," said one participant. "And the part where we looked through the microscope and saw all that stuff, it was fascinating. I could have done that for hours."

Enthusiasm for the Piedmont Project was strong. When asked what they liked best, most participants echoed the person who said, "The chance to learn from a wonderful group of Emory colleagues." "I liked the group, the creativeness of the other people about their courses," said another. "I didn't realize I was going to enjoy the group so much. It was a really big thing for me," added a third.

Why has the Piedmont Project generated such enthusiasm, especially among the majority who had not been involved before with campus environmental issues? There are probably as many reasons as there are participants. A strong format and great facilitators are crucial, as well as the fact that participants' own expertise is affirmed, and they move toward environmental issues from the security of their own specializations. It is also an opportunity to return to our original intellectual curiosity and love of learning. Some like stimulating debate about issues, but probably all appreciate the rare chance to have fun in nature with colleagues—and to get (modestly) paid for it.

The satisfactions of joining our daily educational work with personal values and discovering that those commitments are shared is an important dimension of the Piedmont Project, echoing the gratitude and relief expressed during the mission statement consultations. Said one faculty member, "It matters to me that I sense a certain moral commitment [among the participants]. . . . Everyone who signed up for this workshop believed these things really matter. It let me throw myself into it." Another commented that what stood out for her was "meeting people who were passionate. . . . This was an aspect of [friends] I hadn't known before. It was eye-opening . . . a community of like-minded people." And another said, "Maybe how we identify as a person and as a professional are separate, and maybe with environmental issues they're brought together. Maybe we have a belief in the importance of these issues, so we put aside chasing the resumé, recognizing that 'it's something bigger than you.' " A scientist recalled, "We were all part of a movement, pulling together in something important . . . friendship with action."

The Piedmont Project will probably need another five years to reach a critical mass to embed sustainability issues firmly in the curriculum. For the second and third summers (2002 and 2003), new faculty leaders have come forward to help. With able staff support from Science and Society, my biology colleague and I provide continuity without being overworked. Each new participant broadens campus awareness. Some course revisions have fostered new research directions and professional opportunities, unforeseen at the beginning of course revision. Secure funding is a challenge in a time of budget constraints, but the Green Lunch Group and the Piedmont Project seem to have worked as vehicles to allow engagement in the university's sustainability efforts. Though it took work to get the ball rolling, many faculty now have loyalty to these efforts, and it builds legitimacy for environmental change. The affirmation and enthusiasm of colleagues has also been a source of renewed energy for its leaders.

Conclusion

Each of the environmental activities devised to transform Emory's culture has served to restore some of the intrinsic rewards of the academy: collegial engagement, connection to socially relevant issues, and the intellectual curiosity of the academic life. Building community and establishing, over and over, trusting personal relations is crucial. Emory still lacks a coherent university-wide program of environmental efforts (though the president's task force has now made far-reaching recommendations), but the respectability of advocating for environmental concerns is vastly different now, and many independent green activities are bubbling.

In retrospect, the Ad Hoc Committee built momentum by sweating over ivy, sharing our wonder at mature forests, and working on the mission statement—highly visible early successes. Many of us genuinely enjoyed the diversity of the group, with engineers from facilities management, students with different majors, lawyers, theologians, and administrators with various portfolios, all coming together to share information and try to make a difference. We found, from the library to purchasing, from recycling volunteers to environmental studies faculty, that there is a hunger to connect. We also relished the kinesthetic, experiential learning about the place where we are located. Grounding ourselves in the campus spaces has been a delight.

One component of the success we achieved was the signals of support from the provost, the president, and several vice presidents and deans, who at several pivotal points helped reassure some who feared high-level disapproval of our efforts. Affirmation from the top, joined with massive attention to the grass roots, was important at Emory. Several administrators provided funding at crucial junctures, which reinforced the viability of initiatives. Also important was that over a decade of work by the Senate Committee on the Environment, Facilities Management Division, Alternative Transportation, Recycling, and Purchasing laid the groundwork for campus awareness of willingness to act.

It will take persistence to build on these early steps, and I hope new leaders will continue to emerge. But I no longer worry that having only one person to start the ball rolling is a poor start. I am ready to step back from the limelight, but I am no longer a reluctant leader, and I know that someone has to do this basic administrative and group maintenance work. Someone has to reserve the meeting rooms, facilitate the agenda, send out the notices, and think about the long term. Someone has to create the space in which dialogue can occur and model the trust in the wisdom of the group. I have learned that if we go where the energy is, the group will expand. It has also been a lot of fun.

What are the lessons for me? I have learned that things that seem at first to be a failure may have some later payoff. I also learned to let go of ideas when the energy for them has died. There may actually be a benefit to a nonscientist in a leadership role, in that it decenters the place of the expert and invites broader participation. It certainly keeps me learning more science. I have also learned that it is important to keep articulating why we need to be better stewards of our resources, to reinforce our resolve and to highlight these values for those who have yet to attend to them. This repetition of our purpose builds the need for action, and I have learned that in spite of my shortcomings, I can sometimes nurture a group voice, and I find enormous satisfaction in watching environmental efforts unrelated to the Ad Hoc Committee unfold across the campus.

Where did I find ways to sustain myself as a leader? I went to a lot of workshops, learning from other schools. I took to heart advice from a Second Nature conference to "find five people you really enjoy being with and gather each month." At least twice a year, I take quiet time to get out my list of long-term goals and reflect on how we are doing and what would be the most fruitful next steps. I have not been afraid to use my own money from time to time. I think of it as a support to myself to hire a student helper or skip the hassle of getting a department budget to cover an expense. Also important, I have let in others' affirmation for what we are doing. Friends who choose not to join activities might once have disappointed me. Today, I am grateful for their words of support, and I listen less to the curmudgeons and cynics who believe that nothing can happen. I also try not to feel naive as we celebrate the baby steps. We need to feel the satisfactions of our movements forward. After two years of making campus action my professional work, I began to feel a tug toward research and writing, of which this book is a part. Of course, I am aware that the privileges of tenure make my decision to shift the nature of my work more possible for me than for many others. In the end, only our internal wisdom knows whether we have been good stewards of our "wild and precious life" (Oliver 1992, 94). For me, there is no doubt that the joyful learning and the satisfactions of campus change are worth the costs. Ultimately, we cannot know the results of our actions, and it has been important to me to act with as little attachment to outcome as I can manage. As the ivy has receded, the mission statement adopted, the Piedmont Project established, I am aware that things could have turned out very differently. All we can do is seek the steps that seem wisest. We each contribute our few drops to the flowing river of cultural change.

Acknowledgments

I thank Susan Barlett, JoAn Chace, Geoffrey Chase, Rebecca Chopp, Anne Farber, Howard Frumkin, Kathie Klein, Julie Mayfield, Marc Miller, Mary Elizabeth Moore, Karen Mumford, Bobbi Patterson, Laurie Patton, Debra Rowe, Sonya Salamon, Steven Sanderson, and John Wegner for inspiration, critiques, and suggestions for this chapter.

Notes

1. Parallel to the efforts described here are several major developments in Facilities Management that resulted in three green buildings on campus and the adoption of Leadership in Energy and Environmental Design (LEED) guidelines for campus construction (Wegner 2002). Though I have emphasized here the various efforts to build grassroots groups and to change the Emory ethos, it is possible that Emory's LEED construction has had even more long-term impact to raise environmental awareness because of the widespread media attention it received. Other environmental efforts that cannot be included in this discussion are the founding of Friends of Emory Forest, the development of the Lullwater Management Plan and the No Net Loss forest policy, and liaison with Peavine Watershed Alliance.

2. Baker Woodland is adjacent to the main quadrangle of the undergraduate campus of Emory. The forest affected by the new road is much larger and at the eastern edge of campus, across a major thoroughfare.

3. The walking tour effort was actually part of Millennial Year events, which included a major all-campus workshop ("Nurturing a Green University") led by Second Nature, and conference appearances by David Orr and E. O. Wilson.

4. The Second Nature Southeast Regional workshop stimulated immediate action toward incorporating LEED principles into the Whitehead Medical Research Building, then under construction. This building has been awarded the second level (silver) LEED certification.

References

Barlett, Peggy F. 2002. "The Emory University Walking Tour." *International Journal of Sustainability in Higher Education* 3(2): 105–112.

Barlett, Peggy F., and Arri Eisen. 2002. "The Piedmont Project at Emory University." In *Sustainability in Higher Education*. Edited by Walter Leal Filho. Frankfurt: Peter Lang.

Calder, Wynn, Rick Clugston, and Thomas Rogers. 1999. "Sustainability Assessment at Institutions of Higher Education." *The Declaration* 3(2):1, 17–18.

Capra, Fritjof. 1996. *The Web of Life*. New York: Anchor.

Creighton, Sarah Hammond. 1998. *The Greening of the Ivory Tower.* Cambridge, Mass.: The MIT Press.

Davey, Andy, Graham Earl, and Roland Clift. 1999. "Driving Environmental Strategies with Stakeholder Preferences—A Case Study from the University of Surrey." In *Sustainability and University Life* (pp. 47–66). Edited by Walter Leal Filho. Frankfurt: Peter Lang.

Jaworski, Joseph. 1996. *Synchronicity: The Inner Path of Leadership*. San Francisco: Berrett-Koehler.

Katzenbach, Jon R., and Douglas K. Smith. 1994. *The Wisdom of Teams*. New York: HarperBusiness.

Mumford, Karen G. 2001. "Developing a Campus-Wide Environmental Policy: 'It's All About Process . . . ' " In *Conference Proceedings, Greening of the Campus 4: Moving to the Mainstream* (pp. 267–270). Ball State University.

Oliver, Mary. 1992. *New and Selected Poems*. Boston: Beacon Press.

Senge, Peter M. 1990. *The Fifth Discipline: The Art and Practice of the Learning Organization*. New York: Currency/Doubleday.

Wegner, John. 2002. "Emory University, Building Design." In *Campus Ecology Yearbook, 2001–2*. Washington, D.C.: National Wildlife Federation.

Wheatley, Margaret J. 1992. *Leadership and the New Science: Learning about the Organization from an Orderly Universe*. San Francisco: Berrett-Koehler.

II

Redesigning the Curriculum

4

The Ponderosa Project: Infusing Sustainability in the Curriculum

Geoffrey W. Chase and Paul Rowland

Northern Arizona University is a state-supported, comprehensive university with nearly 20,000 students. Located in Flagstaff, a community with 58,000 residents, the 730-acre campus sits at an elevation of 7,000 feet in the largest contiguous ponderosa pine forest in the United States, just south of the San Francisco Peaks and the Grand Canyon.

Since 1995, more than 100 faculty at Northern Arizona University (NAU) have taken part in a faculty development program called the Ponderosa Project. Coming from areas as far ranging as music, geology, nursing, English, philosophy, engineering, business, political science, education, and art history, these faculty have integrated issues of environmental sustainability into 120 courses across the curriculum. In part because of this project, the new general studies program implemented at the university in 1999 included as one of its chief emphases a focus on the environment. These conversations led in turn to the establishment of the Center for Sustainable Environments in 2000. The Ponderosa Project has provided and continues to provide a forum through which faculty across campus can explore the necessity for interdisciplinary approaches to research and teaching about sustainability while also lobbying the administration for a stronger university-wide commitment to sustainability.

Not surprisingly, the Ponderosa Project is, by most estimates, very successful at NAU and has served as a model for other faculty development projects across the United States. Nevertheless, the seven-year history of the Ponderosa Project is marked by significant peaks and valleys (sometimes canyon-like) in faculty and administrator interest, support, funding, and participation. The year 2000–2001 was a particularly difficult transition time for the university, and the impact on the Ponderosa Project was significant. During that year, the university experienced the end of one president's term, a short and controversial second presidential term, and the appointment of the current president of the university. These transitions were accompanied by a declining resource base, declining student enrollment, and a significant drop in faculty morale. Despite this negative context, the Ponderosa Project pulled itself up by its bootstraps to make the 2001–2002 academic year a success despite the loss of key leadership and the failure to conduct a summer workshop in 2001. This was not the first time that the project had to regroup and reassess its role in the university, and we believe it is important to recognize that even successful projects and efforts inevitably face significant challenges in sustaining themselves. Our aim in this chapter is to describe how the project began, moved forward, stalled, and rejuvenated itself. We also focus on how those of us involved in the project met these challenges—sometimes successfully, sometimes less so—and worked to keep the project moving forward. These are important lessons not only for those of us who were involved with the Ponderosa Project, but also for those at other colleges and universities. How, we might all ask, can we continue to help higher education contribute to a more sustainable future for us all?

Early Involvement: Geoff Chase

In the fall of 1994, while serving as the director of English composition at NAU, I received a call from Henry Hooper, the vice president of research and graduate studies. He asked that I attend a workshop in Miami with Joan Jamieson, an applied linguist, who was at that time NAU's associate dean of arts and sciences. The workshop, organized by Second Nature, was part of a broader initiative funded by the Department of Energy that focused on members of the Historically Black Universities and Colleges and Minority Institutions consortium (HBCU/MI). NAU, because of its substantial Hispanic and Native American student population, is a member of this consortium. I was asked to go because I had rewritten the English composition curriculum to draw on environmental texts representing a broad range of genres. As part of these efforts, I had also produced for the required composition course a custom-published reader, *Critical Reading and Writing in the University Community: The Environment.*

The workshop was designed to help faculty and other administrators become leaders on their campuses and to promote sustainability across the curriculum. During the three-day workshop, we were presented with several case studies that focused on sustainability, considered ways that such materials might be infused into the curriculum, and discussed the important role that higher education has to play in sustainability efforts. The challenge with which we were presented was to return to our own campuses and create programs that would enable a broader group of faculty to include issues of sustainability in their teaching. The three days were packed, and they were both exhausting and inspiring.

When Joan Jamieson and I returned to campus, we began meeting with Paul Rowland, who had a split appointment in the Center for Excellence in Education and the Center for Environmental Sciences and Education; Bill Auberle, the university's first faculty member in environmental engineering; and Susan Schroeder, a grants and contracts officer. Joan and I learned that NAU's involvement with Second Nature, a nongovernmental organization dedicated to involving higher education in sustainability efforts, and HBCU/MI had preceded our participation.

Early Involvement: Paul Rowland

Second Nature, under Tony Cortese's leadership, began working with HBCU/MI in 1992. The HBCU/MI had already established the Environmental Technology Consortium whose chief aim was the revision of curricula to bring more minorities into environmental fields. These efforts focused initially on curriculum change in environmental sciences and engineering. Once Second Nature became involved, the scope of the project broadened to focus on curricular change beyond engineering and the sciences. NAU's involvement began at the formation of the consortium when Henry Hooper argued that the university could provide the Department of Energy (DOE), the funder of the project, access to large numbers of Native American students. DOE was especially interested in the inclusion of NAU as well because many DOE facilities with environmental problems are located in the western United States near tribal lands.

Northern Arizona University was thus one of eight institutions willing to work with Second Nature to explore efforts to green the curriculum. Henry Hooper asked me to be the partnership leader for NAU and I attended two consortia workshops subsequent to the Miami meeting to work through the Second Nature approach. The first workshop was organizational but also demonstrated the model that Second Nature provided to the group. In the second workshop, the partnership leaders (including representatives from a similar Brazilian environmental consortium) conducted a number of sessions on various aspects of sustainability ranging from sustainable agriculture to environmental racism. In the summer of 1994, the HBCU/MI consortium scheduled its annual curriculum meeting at NAU, and all eighteen member institutions sent representatives. Two days were set aside during this meeting to model the Second Nature workshop approach, and along with outside experts such as Robert Bullard, an international expert on environmental racism and professor of sociology at Clark Atlanta University, I participated as a resource expert in the area of fundamental ecological knowledge. It was clear from hallway conversations that this meeting represented a significant shift for the consortium away from funding for the scientists and engineers and toward funding for faculty across the university.

Coming Together on Campus

When all of us who had been involved in these efforts sat down in the winter of 1995 to plan our first workshop, aimed at faculty on our own campus, we came with some common background but different disciplinary perspectives. Joan Jamieson was trained as an applied linguist, Geoff Chase's training was in American literature and composition, Paul Rowland was trained as an environmental educator, and Susan Schroeder, who worked in the Office of Grants and Contracts, had a background in natural resources interpretation. Perhaps significantly, three of us came out of backgrounds—applied linguistics, education, and composition—in which pedagogy plays a key role, and Susan's background involved working with people and was not focused on disciplinary training at the doctoral level. Both Paul and I also had commitments to studying the history of higher education in the United States. In retrospect, the fact that all four of us remember laughing as much as planning suggests that the relationships we built early on were a critical part of our success.

With Henry Hooper's support and with funds he had received through the DOE grant secured by the HBCU/MI consortium, we began planning the first workshop for faculty at NAU. Our aim was to help faculty integrate issues of environmental sustainability into a wide array of courses. We set several explicit goals for ourselves, and these were critical in helping us move forward. First, we decided that in this initial year, we would send out invitations to about forty faculty—our target was to end up with twenty participants—who we thought might be most interested in the aims of the workshop.

Second, we decided that we would require faculty to revise one of the liberal studies, or general education, courses they offered. Our commitment here was to the notion that sustainability represents a broad, interdisciplinary paradigm, one that cannot be adequately addressed in one program or major and must be approached throughout the curriculum. Consequently, in this first year as well as in subsequent years, we invited resource experts who could address sustainability issues from a broad range of perspectives.

Third, we wanted to ensure that sustainability was presented in its broadest aspect, and so we sought to invite outside resource experts who could speak briefly about topics as diverse as economics, social justice, biodiversity, and land use.

Fourth, we agreed that the workshop had to achieve a balance between presenting information and active involvement on the part of the participants. Our sense, even in these early discussions and later confirmed throughout all the workshops we offered, was that faculty would be the real experts in the room and that they would know best how to revise their courses. We believed that these workshops had a secondary purpose of providing a forum in which faculty from across campus could get to know each other better and in which they could become experts and resources for each other.

Finally, we were committed from the outset to workshops that achieved a balance between content and pedagogy. We believed it was essential that the workshops focus on teaching as well as on the acquisition of new content. Our sense was that if faculty were going to change their courses, they would have to focus as much on how they taught as on what they taught.

Several other factors played a large part in our early discussions. The money received from the DOE allowed us to provide incentives of $1,000 for faculty once they submitted a revised syllabus, to hold the workshop at a location off campus, and to invite four outside resource experts who could work with the participants over the two full days of the workshop. These incentives were critical. They indicated to faculty that the upper administration fully supported this venture and that they were willing to provide financial backing to carry the project forward. Like many other faculty members, those at NAU believed that administrators should reward faculty for what they perceived as extra work. In our experience, this perception is widely held, and thus we cannot overlook the power of financial rewards as we work to promote change within the academy.

The First Workshop

The first workshop we offered, in 1995, set the pattern for all those that followed. On the evening before the workshop began, those of us who had been planning the workshop met for dinner and conversation with the resource experts we had invited. This provided the four of us with an opportunity to talk about the goals of the workshop and to describe more specifically the roles we hoped the experts would play. We wanted to be sure that they were there as experts but also as facilitators. This evening gathering also gave the outside resource experts an opportunity to get to know each other, to know us, and the opportunity to ask questions.

The schedule for the first day of the workshop alternated between presentation and discussion, with the weight of the time going to discussion. We began with an overview of sustainability, moved to a presentation by one of the resource experts, and then broke for discussion. Over the rest of the day, we maintained the same pattern. Throughout the first day, our primary goal was to provide some information that would open up sustainability as a concept and provide as many openings to the topic as possible. Our second goal was to have the faculty participants get to know each other and talk with each other about their academic disciplines and about the courses they were planning to revise.

On the second day, we began with a presentation about interdisciplinary teaching, and then devoted most of the day to discussions in which faculty worked with the resource experts and each other to talk about how they planned to make revisions to their courses.

At the conclusion of the two-day workshop, faculty departed with the task of revising their syllabi over the summer. During this time, they were free to conduct research, do more reading, and consider questions of content and pedagogy as they revised their syllabi. In August, just before the start of the academic year, faculty who had taken part in the workshops in May reconvened, without resource experts, to share with each other the work they had done and to present their syllabi and courses to each other.

The response from the faculty—drawn from such diverse disciplines as biology, philosophy, history, political science, Spanish, education, hotel and restaurant management, psychology, business, Navajo, and English—was overwhelmingly positive both at the conclusion of the two-day workshop and when we reconvened in August. Faculty spoke highly of the resource experts, but their highest praise was for each other and for the opportunity to move outside their own disciplinary boundaries to work with other faculty across campus. Of the twenty faculty who enrolled in the workshops, eighteen turned in revised syllabi, and a small number of faculty continued to meet throughout the year to discuss the progress of their courses.

The Next Five Years

The underlying assumptions driving all of the workshops we offered over the next five years remained the same: (1) faculty benefit most from being presented with a broad range of approaches, ideas, and resources; (2) education for sustainability is linked to content and pedagogy—how we teach is as important as what we teach; (3) faculty themselves know best how to revise the courses they teach; and (4) one way to help faculty move toward sustainability is to provide opportunities for them to step outside the boundaries of their disciplines and departments, talk to each other, share ideas and insights, and see themselves as essential participants in a larger project.

The changes that faculty made in their courses represented a broad range of possibilities. Some faculty used environmental issues as examples in their courses to illustrate key concepts or ideas within their disciplines, while others introduced environmental content directly. Others consciously changed their orientation so that students are regularly reminded that the material they are addressing is linked to visions of sustainability, and, finally, some created class or individual projects that address environmental concerns. Most often, faculty chose a variety of strategies, so that the courses shifted on more than one level at a time. Three examples provide a clearer picture of these creative changes.

"Medieval Art," an upper division course, examines Western European painting, sculpture, and architecture from c. 350 to c. 1350. The revised version of this course includes medieval and contemporary readings that focus "on the relationship of western medieval culture to its environment and [which] have been selected to highlight the ways in which western medieval culture manifested an awareness of and a connection to the natural world." Some of these readings—nature poetry of St. Francis, bestiaries, herbals, and "health handbooks"—demonstrate an acute awareness of the ways in which plants and herbs could be used to treat common afflictions. In other parts of the course, students are led to consider early efforts at reforestation in thirteenth-century France, which came in response to rapidly diminishing resources as wood was being harvested to support the production of stained glass windows, an art form that required extended firing at high temperatures.

The course also draws attention to the threats that modern pollution poses to the art and architectural monuments that are the legacy of medieval culture. Finally, students are asked to make explicit connections between the medieval world and contemporary society through a reading of *The Temptation of St. Ed & Brother S* (Bergon 1993). In Bergon's novel, the spiritual life of St. Ed and his followers is threatened by the installation of a nuclear waste storage facility being built near the site of the monastic community they have established in the desert outside Las Vegas. The goal of this assignment is for students to understand the continued prevalence of medieval Christian institutions and values in contemporary society (and literature) and to focus their attention on a fundamental environmental issue in ways that are particularly relevant to Southwest culture and geography.

In "Introduction to Archaeology," a sophomore course, students focus on the Black Mesa Archaeological Project in northeastern Arizona as a way of beginning to understand the goals, aims, methods, and theories that shape archaeological research. The course has been revised so that the Black Mesa Project has become an opportunity to raise issues such as sustainability, environmental racism, environmental degradation, and overpopulation. In addition to attending lectures, "students work in small groups to discuss the following suppositions: (1) substances from the earth's crust must not systematically increase in nature; (2) substances produced by society must not increase in nature; (3) the physical basis for productivity and diversity of nature must not be systematically degraded; and (4) there must be just and efficient use of energy and other resources (Erikson and Karl-Henrik 1991, Hawken 1995)."

Students in the course write a paper in which they examine how prehistoric humans may have perceived their relationship to nature and to the environment, and in other discussions they address the impact of mining carried out by Peabody Coal on Navajo and Hopi Indians. In addition, students in the course examine how prehistoric humans modified the environment in which they lived and contemplate how those modifications might have been connected to overpopulation and environmental degradation. Finally, this revision includes not only the addition of new material, but also new teaching methods. Whereas the primary pedagogy used to be lecture, students now engage in in-depth discussions as a way of learning to think critically, to evaluate a range of opinions and issues, and to consider, through a study of the past, their roles in a sustainable future.

Traditionally, little emphasis has been placed on helping students become aware of environmental issues related to organic chemistry. In "Organic Chemistry," however, environmental examples are used to demonstrate some of the key concepts and ideas of organic chemistry where relevant. For example, students focus in lectures and through additional readings on several topics related to environmental sustainability: (1) the disposal of organic waste generated by industry, academic institutions, and privately; (2) the manufacture and use of fertilizers and pesticides; (3) remediation of heavy metals from water and soil; (4) clean-up and disposal of petroleum spills on land and in the ocean; and (5) the design and manufacture of organic materials and alternative sources of fuel, and for collecting and converting solar energy to electricity. Although the course is designed primarily to provide students with the fundamental chemical concepts and principles governing the reactivity and physical properties of organic molecules, students also come into contact with issues of environmental sustainability.

These courses typify the kinds of changes most faculty have made in their teaching as a result of the Ponderosa workshops. These faculty have found that examining issues of sustainability does not mean giving up vital content. Rather, focusing on these issues is a way of making material immediate and relevant to student experience. The interdisciplinary and systems-oriented approaches in these courses have been ways of helping students connect with the material and to understand that their learning is connected to the larger problems they face as students and that they will face as citizens throughout their lives.

Dealing with the Canyons as Well as the Peaks

In spite of the considerable success the project achieved, it also faced key challenges over the five years. After the first year of the project, some of the faculty who had been participants in the first workshop began to take on roles as we planned subsequent workshops, and this pattern continued throughout the five years. The core group, including Geoff Chase, Paul Rowland, and Susan Schroeder, remained the principal organizing group for the workshops.

One of the divisions that arose was between those who saw the Ponderosa Project as consisting of those who had been through the workshop and others who saw it as a broader group that could include anyone interested in environmental sustainability. For the most part, those of us who had designed the first workshop fell into the camp that saw the project as one leading specifically to curricular transformation and not as a broader environmental interest group. This tension remained a challenge throughout the duration of the project. We now realize that this conflict was a function of the newness of sustainability to the university and, consequently, the lack of a variety of forums for discussion of the different approaches to sustainability. Thus, what we saw as an implied request to change the focus of the workshops was in fact a desire for broad campus involvement around a series of environmental issues. Over time, this issue eventually resolved itself as more venues were developed for faculty to participate in green building and green outreach efforts facilitated by the Center for Sustainable Environments.

A second challenge arose around resource issues. Money from the DOE was reduced after the first year of the workshop, and we began to scale back the amount we expended on the workshops. By the third year, our budget was half of what it had been in the first year, and we made up most of the difference by scaling back faculty stipends from $1,000 to

$500 and cutting some overhead costs. We also found ourselves working with new administrators—Henry Hooper stepped down from the vice presidency in 1998—who, although supportive, had a difficult time finding money to support the project. By the time we offered the fourth workshop in 1998, we had run out of funds altogether and decided not to offer the project in 1999. When we did offer the fifth workshop in 2000, we funded our efforts partly by cost savings—we did not offer any stipends to faculty at NAU—and by charging a small fee to faculty members who wished to come from other campuses. Because of a lack of funds and continuing changes in university personnel, the workshop was not offered in 2001.

Funding for a project such as this is particularly difficult because it is part of no university budget unit and reports to no one administrator. As we noted, the initial funding from DOE enabled us to demonstrate the support for the project from upper administration. But since this funding was soft, we did not have a good way to sustain the project financially over time. Some faculty who came into the project later understood but were also frustrated by the fact that we had to halve the stipend we offered. Consequently, we believe it is essential eventually to move away from perceiving curricular transformation around issues of sustainability, or other interdisciplinary initiatives for that matter, as extra work that can occur only when outside funding is available. Given the rigidity of departmental boundaries and the fact that ways of thinking critical to achieving a more sustainable future for our world are fundamentally interdisciplinary, we need to find a way to provide ongoing support from within our institutions for these efforts.

A third challenge arose as we reconsidered the aims of the project over the first five years of its existence. Initially, the workshops focused exclusively on helping faculty revise what they taught in individual courses. By the third year of the project, we shifted the agenda of the second day of the retreat to include a discussion of what students would *learn* as a result of those revisions. This shift in focus on what is taught to what is learned was consonant with assessment efforts across the university and reflected what has become a national dialogue about the role of higher education in our society.

Such a shift was important, for it led us to reshape the workshops more consciously around curricular issues broadly understood. At the same time we began asking faculty to articulate what students would learn in their courses, we began to ask them to articulate what students should learn about sustainability through the whole curriculum.

A fourth challenge emerged when Geoff Chase left NAU in fall 2001. Geoff had provided key leadership throughout the project and had been able to use his various administrative offices to provide support for organizational activities. It is important to note that the Ponderosa Project, like any other project, needs a leadership node that will call and organize meetings, facilitate activities, and ensure follow-up. Geoff and his office staff had played this critical role, and his departure threatened the continuation of the project.

Fortunately, a group of faculty who had participated in the workshops felt that the Ponderosa Project was too important to let it die, and in the fall of 2001, they formed the Ponderosa Project Steering Committee. About fifteen faculty and staff members joined the steering committee and began looking for opportunities to revitalize the project. The group decided that there were five key activities in which they should engage to ensure the infusion of sustainability into the curriculum would continue. First, they planned a campus-wide reception for faculty and administrators. Second, they planned a half-day rejuvenation workshop. Third, they began looking for funding for a regular summer workshop. Fourth, faculty met with the spring 2002 instructors of the "University Colloquium" course, a seminar required of all incoming first-year students, to explain the role of Ponderosa Project and to invite their participation. Finally, the steering committee began exploring other avenues to influence faculty understanding of issues of sustainability.

A reception for the campus community was held by the Ponderosa Project in January 2002. The featured speakers at the reception were President John Haeger, Paul Rowland, at that time chair of environmental sciences, and Marcus Ford, a participant in the first summer workshop and a member of the steering committee. About eighty faculty members attended this reception and received the message that education for sustainability was still alive at NAU and that they were welcome to participate.

Abe Springer, a geology professor who became involved in the Ponderosa Project as a participant and later as a resource person, organized the half-day rejuvenation/reunion workshop in March for faculty who had participated in the Ponderosa Project. As part of this workshop, faculty were given an opportunity to give input into the planning process for a new green building for applied environmental research that will soon be constructed on the campus.

In the later fall of 2001, the Arizona Board of Regents announced a funding initiative to support learner-centered education (LCE). The Ponderosa Steering Committee agreed that it was the perfect place to seek funding and submitted, through the NAU Faculty Development Office, a proposal to extend the Ponderosa Project summer workshop to a tri-university workshop with significant follow-up. When the proposal was reviewed, it was given high marks, along with a proposal from the University of Arizona, and the three faculty development offices were charged with putting the two proposals together. The result was that a workshop was held at NAU for faculty from the University of Arizona, Arizona State, and NAU, and that workshop included education for sustainability as a theme. Marcus Ford served as the organizer for those sections of the workshop. Interestingly, a theme that had been emerging in recent Ponderosa Project workshops, assessment of student learning, was also the primary topic advocated by the University of Arizona faculty development office.

Recent subsequent developments have also helped sustainability efforts gain a more prominent position in the university community. The emergence of Marcus Ford as a leader in this effort coincided with his election to vice chair of the faculty senate. As such, he obtained a seat on the NAU Strategic Planning Council. Also, Paul Rowland moved into a new campus position as director of academic assessment. He also was assigned a seat on the Strategic Planning Council. As that body began considering clarifying the future directions of the university, Marcus and Paul decided that a campus-wide push for recognizing the significance of environmental sustainability at NAU was due. In early fall 2002, in order to demonstrate the widespread support for a greater emphasis on environmental work at NAU, they organized a meeting entitled "NAU—The Environmental University." Although university presidents had used this tag line in the past, it appeared that the tag was being lost in various university documents. The power of the Ponderosa Project to pervade the university was evident as more than 125 faculty members attended the meeting and spoke about the importance of environmental sustainability to students, faculty, and the university as a whole. This meeting, reminiscent of one nearly a decade earlier, brought together faculty engaged in Ponderosa-style sustainability education, environmental researchers, and environmental outreach faculty and staff. Although it is unlikely that one will see new brochures extolling “NAU—The Environmental University” anytime soon, there has been a clear increase in the use of *sustainability* and *environment* in university planning documents. The most important point of this meeting is that despite our ongoing efforts, it is important periodically to reaffirm who we are and what we are doing for the administration and for each other.

Lessons Learned

The past several years have taught us many lessons about the process of increasing education for sustainability on a university campus. First, it is important that the process have support. The Ponderosa Project would not have been initiated without the support of a key administrator, associate vice president Henry Hooper. He was able to identify and bring together the key people who started the project. His support was critical in a second realm: funding. New projects can be started with lots of sweat equity, but they are difficult to continue unless some source of funds is located early in the process to show that the institution values the efforts.

A third lesson was the need for leadership. It was critical for the Ponderosa Project that it have a well-bonded team of four leaders, but that it also open the door for emerging new leaders. Although the Ponderosa Project has been praised for its flatness and grassroots structure, there is a persistent need for leadership. Its continuing success is the result of being not too centralized and yet maintaining some level of leadership. Finally, a key element in the continued success of this project has been a clarity of focus. The ability of the Ponderosa Project to avoid becoming all things for all people has been essential in its continuation. Although it has evolved significantly over time, it has avoided becoming “the” environmental organization on campus and has allowed other groups to emerge and play critical roles. By keeping its focus on education for sustainability through faculty development, it has been a key player in developing a university-wide commitment to provide leadership in building a sustainable future.

References

Bergon, Frank. 1993. *The Temptations of St. Ed & Brother S.* Reno, Nev.: University of Nevada Press.

Chase, Geoffrey. 1993. *Critical Reading and Writing in the University Community: The Environment.* Fort Worth, Tex.: Harcourt Brace Custom Publishers.

Erikson, Karl, and R. Karl-Henrik. 1991. "From Big Bang to Sustainable Societies." *Review in Oncology* 4(2):5–14.

Hawken, Paul. 1995. "Taking the Natural Step." *Context: A Journal of Hope, Sustainability, and Change* 41:36–38.

5
Transdisciplinary Shared Learning

Richard B. Norgaard

The University of California at Berkeley is the oldest of nine campuses of the state-funded University of California system. It occupies an urban 1,200-acre campus at the foot of the Berkeley hills, across the bay from San Francisco. Berkeley has an enrollment of 22,700 undergraduates and 8,800 graduates and offers more than 300 degree programs.

The Energy and Resources Group (ERG), an interdisciplinary graduate program at the University of California at Berkeley, was initiated through a process of shared learning among scholars of multiple disciplines, and this evolved into a central sustaining characteristic. In 1969, C. K. (Ned) Birdsall, a plasma physicist in the Department of Electrical Engineering and Computer Sciences, initiated a seminar series broadly addressing the resource bases, technical requirements, environmental consequences, and economic and social implications of alternative energy futures. Being centered on shared learning rather than a particular approach to environmental synthesis has kept ERG among the leaders of innovative, interdisciplinary thinking for three decades.

ERG is unique among the many graduate groups at Berkeley because it adapted as the domestic environmental and energy crises evolved into the global environmental, cultural, and justice issues of sustainable development. Now, with three decades of unfolding success, we can look back and see how fortunate our beginnings were, as well as see how some of the steps taken along the way proved crucial.

The seminar Ned Birdsall initiated became an established feature on Wednesday in the late afternoon, attracting more faculty and a broader array of participants over the early 1970s. A brown bag lunch with the seminar speaker was added to the schedule, providing an informal opportunity to discuss current energy developments as well as how the campus might respond. Physicists dominated, and sometimes overwhelmed, the effort, in part because of the proximity of researchers from the Lawrence Berkeley National Laboratory, an Atomic Energy Commission (soon to become Department of Energy) research laboratory adjacent to the campus. The common mind-set and language of the physicists fortunately was matched by an uncommon dedication to the larger questions. They courted the voices of faculty from biology, law, political science, public policy, city and regional planning, and anthropology.

I joined the group in the fall of 1970 as a young assistant professor in agricultural and resource economics. I had grown up in the area, done my undergraduate work at Berkeley from 1961 to 1965, and become embedded in the West Coast–based environmental movement as a river guide to David Brower then executive director of the Sierra Club, and others in the Glen Canyon of the Colorado River before it was inundated by Lake Powell. I had spent the summer of 1970 studying the environmental conflicts of Prudhoe Bay and the TransAlaska Pipeline. For my doctoral dissertation in economics at the University of Chicago, I had done research on petroleum scarcity and well-drilling technology. Thus, I had information and insights to share.

Interdisciplinary communication worked because we were pragmatic empiricists who felt a curious need to know how things—natural, technical, and social—work. Our need to know was also driven by a sense that we could affect the future. The social scientists among us were then more positivist than we are now, which eased communication with the natural scientists. And Ned Birdsall set the tone by quietly posing questions about the social, economic, and environmental implications of different energy technologies. They were good questions respectfully framed, well cleansed of implicit challenges too tightly tied to a prior analytical frame or personal ego. We responded from our different disciplinary perspectives and experiential bases and then a week later found ourselves asking more difficult follow-up questions. Social and environmental scientists in turn asked about alternative energy technologies in greater detail. This shared learning process helped each of us to dispense with our simple, and sometimes wholly false, assumptions about the broader reality that had evolved within our isolated disciplinary communities. The group attracted scholars prone to poke their heads out of the comfortable shells of their academic departments. Whole new realms of questions opened up, most of them beyond the territory claimed by any of the disciplines.

Working together across disciplines proved an effective way to explore ideas, triangulate points, and steady our footing. In the unsettled times leading into and through the energy crisis, the boundaries between academe and policy making were readily traversed. Poignant observations came easily; modest amounts of data and a little back-of-the-envelope reasoning proved powerful. And this activist-scholar spirit, process of sharing, and analytical style infused the new graduate program we started. This shared learning process, established at the very beginning among some twenty-five faculty, proved critical to our early success and set a pattern for the graduate program to be.

By 1973, we envisioned a program that would both sustain faculty interaction and train graduate students. Our own shared learning experience documented the possibility for a graduate program to train a new type of professional: someone who could ask critical questions about possible energy futures across a range of disciplines, help put answers together in a practical way, change the way others thought, and help transform institutionalized decision making in the process. We did not think of the training as interdisciplinary in the more conventional sense of filling predefined voids of critical knowledge between a few disciplines. Rather, we saw energy and other societal problems as dynamically emerging through broad systemic interactions. The disciplines helped provide conceptual models of the economic, social, ecological, and technological systems. Thus, our model was to prepare students to ask the new, broader questions while providing them with sufficient disciplinary understanding to pursue systemic answers working with disciplinary experts.

We saw ERG as a professional master's degree for most students, with the expectation that a few students might go on for a more academic Ph.D. Yet our philosophy of education was anything but professional. We had neither a tradition to follow nor a profession to fill. The attitude we instilled was one of curious boldness rather than of staid professional discipline. Things evolved very well from these unusual beginnings.

Program Implementation and Early Successes

Luck was on our side early on. Mark Christensen (geology), an active member of our group, ascended to the position of vice chancellor and gave the group its first official status as the Energy and Resources Committee. He accepted and almost single-handedly pushed, backed, and implemented the argument that our proposed program, unlike any campus-wide research initiative or graduate group heretofore, needed a full-time faculty member assigned to the group to sustain and help integrate the interactive learning between the faculty dispersed among the departments. Having faculty within the group was truly an anomaly, for it meant the campus would have faculty for the first time who were campus-wide and not in a department or under the administration of a dean of a college.

John P. Holdren became the first professor in ERG. Trained in aeronautics/astronautics and plasma physics at MIT and Stanford, Holdren had explored environmental questions as a graduate student working on the side with Paul Ehrlich as well as subsequently serving as a senior research fellow at Caltech. Becoming the first professor of energy and resources proved mutually advantageous for him and our program. Holdren energized our shared learning. He worked very effectively across the disciplinary cultures of academe, the institutional cultures of the Lawrence Berkeley and Livermore Laboratories, and the political and bureaucratic cultures of Washington, D.C. Following Birdsall's lead, Holdren listened and learned, helping all to piece the dynamic processes together, while also helping all to keep moving effectively to the next question. Together, Birdsall and Holdren, with solid, broad support from the faculty in the group, began to implement a program that had yet to be academically approved.

Our application for a graduate program was off to the first of nine levels of academic, administrative, and statewide council review and going nowhere fast when the Arab oil embargo was imposed in late 1973. Vice Chancellor Christensen strategically used the chaos of the energy panic to establish the program administratively in advance of academic approval, but academic approval, though still eighteen months to come, was also greatly accelerated by the crisis. Thus, in a short time, over the next few years, John Harte, a particle-physicist-turned-ecologist, Anthony C. Fisher, a natural resource economist, and Gene Rochlin, a particle-physicist-turned-political-scientist, formed, along with Holdren, what became known as the core faculty.

Due to the prominence of the energy crisis in the beginning and a strong track record since, ERG has been able to attract excellent students. We deliberately sought students from a broad range of majors, many of whom had spent several years after obtaining their undergraduate degree actively involved in social and environmental issues. Core courses on energy and society and environmental system analysis were designed for all students. We then required the students to go through different disciplinary hurdles. The range of academic backgrounds and professional experience of the students has meant they have been in a position to teach each other, much as the faculty had been doing for several years. The students then take on an individual master's project of their choice while staying in touch with what their fellow students are doing through seminars and study groups. Most have soon discovered that their research questions are interrelated, that the difficulties of doing research in different areas have similar epistemological roots, and that they can help one another. Students have been interested in each other's projects because we have selected students concerned about the big picture.

While engineering schools at the time were emphasizing conventional technologies and resources, ERG faculty and students were actively pursuing the more novel possibilities for energy efficiency. Our faculty and students documented national differences in energy use, engaged in home audits, and helped the Public Utilities Commission see that a kilowatt-hour saved is just as good as a kilowatt-hour produced. We elaborated the environmental and social costs of pursuing historic energy paths and helped identify renewable alternatives. We also did key work on acid rain, nuclear security, the social requirements of managing nuclear power, and the possibilities for decentralized power systems. Our research on specific issues was always linked to the larger context, including possible contexts to come. This set us apart from other energy programs.

Our approach proved successful. After two years interacting with faculty and students in our graduate program, our master's alumni with prior training in engineering were able to pose questions in economic terminology that would numb a conventionally trained neoclassical economist at the public utilities commission, opening up possibilities for energy efficiency to be treated on a par with energy supply in the regulatory process. Graduates with prior training in economics were able to ask sophisticated environmental questions, and graduates with prior training in environmental science were able to dumbfound engineers. Using deeper questions to keep those with conventional training in the disciplines and professions off-guard offered possibilities for pursuing new solutions that addressed the interconnected nature of the problems.

The physicists from the Lawrence Berkeley Laboratory (LBL) interacting in our group established an Energy and Environment Division. These developments at LBL were critically important to ERG's success. The lab was able to bring in substantial new research funds to study the implications of alternative energy paths, especially the possibilities for energy conservation. ERG provided bright, hard-working students, faculty with creative and critical insights, and an array of talents before whom ideas generated at the lab could be tested. LBL funding for students was very helpful, and a significant number of students subsequently found employment at the lab. With the lab and campus programs working together, the research output was greater and stronger than it would have been if the units were isolated.

California responded to the energy crises of the 1970s more aggressively than other states. The university established an institute to fund energy research, much of which went to ERG in its early years. The state established an Energy Commission to scope out alternative futures, and ERG graduates assumed key positions early. The Public Utilities Commission responded with new policy initiatives and procedures to promote energy efficiency and renewable energy technologies, and ERG alumni were the ideal job applicants to shoulder these new efforts. There was no better state in which to be an activist energy scholar, a student learning about the possibilities for change, or an alumnus or alumna prepared to take on a challenging job.

In retrospect, ERG took off rapidly into sustained flight because of the energy crisis, but the direction and altitude at which we flew depended on the good work of key people, as well as the status of Berkeley as a place to do graduate work. ERG benefited immensely during its first half-dozen years from the interdisciplinary discursive process and interest in the big picture of its founders. The cast of characters, the conversations we held, and how this carried over into the process of attracting and training graduate students proved exceptional. It was a very strong start, but the stage was set in the 1970s in a way it has not been quite set since for asking new questions, doing back-of-the-envelope calculations grounded in a few data points, and speaking out publicly. The times changed, and our style of asking new, challenging questions also meant ERG was destined to keep moving into new territory.

Broadening Our Mandate

And move we did from domestic energy issues to international ones. Stepping out of our own cultural and institutional base in the 1980s, many in ERG increasingly realized that there are no energy and resource problems, only people problems. Students with prior training in the natural sciences and engineering began to use our program as a transition into the social sciences. Geologists became political scientists and physicists transformed into sociologists, without having their initial footing in the natural sciences kicked out from under them. We clearly became a graduate program that proved an attractive place to pursue an academic Ph.D. Our enrollment grew from twenty-five master's students and a few Ph.D.s to twenty-five of each during the 1980s. Now the stock of Ph.D. students numbers thirty-five, though the flow of master's students is still greater. This transition occurred steadily, yet in hindsight, it was a revolution.

In the 1970s, faculty and students were on a long tether, pursuing environmental, economic, and equity issues, but the tether was firmly anchored to the energy crisis. The crisis unfolded to an economic tragedy for developing countries in the early 1980s that blended with new concerns: biodiversity loss, climate change, the possibilities for sustainable development, environmental justice, and the role of local communities and indigenous voices. The research interests of faculty and students spread out and, for most participants, became quite detached from energy. Our name was Energy *and* Resources because we foresaw other material-driven systemic crises. We had envisioned other anchors in the material world, but we had not envisioned parallel anchoring in the social sciences to understand crises as driven by social factors and dynamics.

With this shift in perspective, ERG has made novel contributions to thinking about social and environmental systems. Our students did some of the best early conceptual and empirical research on environmental injustice. With good backgrounds in environmental science and engineering, they have contributed to environmental history, political ecology, and the social studies of science and technology. Unencumbered by disciplinary culture, they have pushed neoclassical reasoning where economists feared to go. Much of my own work during the 1980s shifted toward understanding social and environmental change as a process of coevolution among knowledge, value, organizational, technological, and environmental systems. And from there I moved into the even broader abstractions of epistemology, of how environmental challenges raised new issues with respect to how we know. Thus, I began to argue that modern ways of perceiving the world and beliefs about how we think we know it, and how we then organize ourselves around these beliefs, underlie our energy and environmental problems and questions of sustainability in general.

The transition was steady but never easy. It should be understood as a broadening rather than as a shift into the social sciences, for at the same time on the environmental science front, John Harte and his students made significant contributions to our understanding of biodiversity and climate change, and these parallel advances have been sustained since. The energy issues that focused campus-wide academic conversations during the 1970s became too dispersed to draw the heretofore faithful founding faculty to our Wednesday colloquia on a regular basis. But many new faculty were being recruited, so that the total number of faculty in the group steadily grew to over 100, although different faculty would appear for different seminars. As our social inquiry shifted from measures of energy demand elasticities and environmental valuation to the historical roots of materialism and modern values, we attracted a new set of social scientists who were critical of positivism and spoke yet more difficult languages.

Another important factor in our transition was that John Holdren continued to be increasingly active and successful as a public scientist: as a leader of Pugwash, the Federation of American Scientists, of National Academy studies, and of science committees reporting to President Clinton. Others of us have aged into greater professional and public service, though nowhere near to the extent Holdren has. He began to spend most of his time on the East Coast, leading him to move to Harvard's Kennedy School in the mid-1990s.

The combination of increased breadth in the topics we were pursuing and increased faculty activity beyond the campus made it far more difficult to sustain an integrating, campus-wide conversation. Though faculty remain loyal to the program for a variety of reasons, including access to exceptional and challenging students, the integrating processes that remain now increasingly occur among the Ph.D. students and core faculty, largely in the Ph.D. seminar and student-initiated study groups. Our Ph.D. program has been very successful from an academic perspective. Over the past five years, ERG Ph.D.s have taken academic positions at strong universities, including Berkeley, in the areas of environmental justice, urban ecology, environmental history, environmental geophysics, ecological economics, terrestrial ecology, alternative energy technology, environmental policy, theoretical ecology, and political ecology.

Another critical challenge is dealing with a campus administration that changes frequently and must constantly be reinformed of our success. The energy crisis is in the distant past, and it is not easy to convey how our program helps faculty and students understand broad problems systemically. Of course, we also have failed to convey effectively how we really operated, too often taking the short-cut of portraying ourselves in interdisciplinary clichés that administrators might understand or relying on the reputations of particular faculty. The administration has initiated academic reviews of ERG approximately every five years. A key question of every review has always been an administrative one: "Why is ERG a campus-wide graduate program, complete with its own faculty, rather than a department in a college with a clearly defined mission duly administered by an existing dean?" And there were college deans on the campus who wanted our strengths, or at least our faculty positions. The College of Engineering, rather than hiring technology and society faculty of its own as other schools do, has periodically eyed ERG as an obvious acquisition. The College of Natural Resources, during an unending period of internal reorganization, occasionally looked at ERG as a natural subcomponent. By subsuming ERG, Berkeley's Graduate School of Public Policy would better compete with Harvard's Kennedy School. Fortunately, with multiple deans fighting "for" us within the memory spans of campus administrators, along with the support of faculty within our group from across the multiple colleges, ERG has remained independent.

Critical Lessons

ERG survived and excelled in new ways for a variety of reasons, allowing our understanding of environment and development to evolve and to resist the challenges of changing campus administrations.

First and foremost, very early in ERG's history we established a national reputation for recruiting, embracing, and enhancing the strengths of very bright, experienced students. Our graduates were actively making a difference in a variety of arenas using empirically based, creative arguments. Now holding increasingly visible positions, our former students proved instrumental in inspiring new applicants. Thus we have continued to have a very strong pool of applicants from which we have been able to select an exciting, complementary mix of students each year.

Second, the tradition of shared learning, especially among the Ph.D. students, continued through the transition. Working on a broader array of topics meant students were taking fewer classes together. This reduced the shared learning among our master's students whose two years in the program emphasizes course work. But the process has been sustained for core faculty and Ph.D. students, who, now being a much larger group, formed their own critical mass through a weekly seminar. Thus even though students have had to choose among numerous classes and select a research project specific enough to be doable, they continued to be engaged with each other in a very broad discourse on the evolving dynamics of the human condition.

Third, ERG has not lost the academic support of disciplinary faculty and departments. Faculty in many environmental programs critique the disciplines and argue that they have collectively developed an alternative approach that is superior. We have tended to emphasize how each discipline helps us see the whole. We have avoided espousing one ERG approach, a single methodology for synthetic understanding, that disciplinary academics might find questionable and an easy target. To be sure, particular faculty, especially me, have not fit into their original disciplinary cultures and have openly critiqued the assumptions and values of our disciplinary peers. But we have distinguished these cultural aspects sufficiently from the methods and knowledge of the disciplines to keep them apart. And even I, along with other core faculty, still publish on occasion in my original discipline. Thus, ERG as a program has maintained respect for and respect from the disciplines.

Fourth, many of the students take some of the most difficult courses on the campus—courses unrelated to their undergraduate training. Many of our students not only have done very well in these courses but have also asked the most creative, challenging questions in class. This ensured that professors across the campus would continue to let our students into their courses, would go out of their way to advise and work with our students on their research, and would continue to keep sustainability issues in mind. This emphasis on crossing disciplines had a surprising effect. ERG became recognized as a place where students could successfully move from physics into sociology, and some have. More generally, students have moved from the natural sciences and engineering into social sciences and policy. One, however, went from English literature and a Yale Divinity School degree into engineering. These students retain their initial training, have done their graduate work with students asking a broad range of questions, and yet also learn how to adapt to a new disciplinary culture.

Fifth, our program has had transdisciplinary faculty in the core: physicists who moved into nuclear security, terrestrial ecology, and alternative energy development and social scientists who learned their natural science. Thus, the core faculty has not been a collection of disciplinary scholars collectively telling our students to broaden out while we individually pursued a narrow disciplinary course. Rather, we have been in a position to empathize with the joys and difficulties our students experience as they learn across the disciplines. The core faculty are able to embrace new ways of thinking while serving as mentors to our students.

Sixth, students have largely initiated their own research rather than plugging themselves into the research of their professors. ERG has been more like a social science or humanity graduate program where funding is perpetually a problem, but students are more likely pursuing their own ideas. ERG attracts risk takers; they live leanly, but they also are participating in defining the program through their personal research.

Seventh, we have sustained the support of our affiliate faculty from around the campus. One of the wisest decisions made early in the program was to allocate 0.33 of a faculty position for the rotating chair to make it easier for departments to lend us a faculty to serve as chair.

Eighth, many of us in the core faculty have also sustained our active involvement in the political and policy processes of worldly issues. We initiate policy statements, testify before legislative bodies, and serve on science advisory boards of agencies. I, for example, see my current service as the past president of the International Society for Ecological Economics, on the Science Advisory Board of the U.S. EPA, as a member of the board of the American Institute of Biological Sciences, and as a member of the boards of Redefining Progress and EcoEquity, and as a participant in numerous other nongovernmental organizations as part of my role in ERG. We also continue to deliberately select students for admission who are activists, but activists who enjoy asking intellectual questions. And we support the continued political involvement of our students while they are in the program. Of course, our activism, like our scholarship, is not as focused as it was during our early, energy crisis years. When challenged as to how one can be an academic and an activist, we respond that not being an activist is a political statement too—an endorsement of the status quo and all of its problems. We question the system rather than merely try to make a few things better within it because our energy, environmental, and social systems are systemically going the wrong way.

The Future

The eight summary statements reflect my weighing of why ERG has had a successful past. I think they can inspire and guide other programs. At the same time, they provide a basis for peering into ERG's future. Recent developments suggest we are on another course. New core faculty have not had the start-up conditions of the original faculty largely because there is not so sharp a crisis focusing a multidisciplinary discourse within which these new careers are being cultivated. We are all more sophisticated with respect to the issues of the past than we were in the 1970s. This sophistication, however, has not translated into general transdisciplinary competence and confidence. Indeed, to some extent, it has led to a greater comfort with disciplinary cultures because they have evolved as history unfolded as well, incorporating energy, environment, and sustainability as respectable, though still marginal, areas of work. At the same time, the tradition-bound disciplines are in no better position to ask wholly new questions than they were three decades ago.

There are plenty of new problems that cut across the disciplines on which we could be working, but words like *energy* and *environment* do not bind the problems together to provide a basis for academic organization. Sustainability rallied our effort for a while, but now it has been used by too many interests in too many contradictory ways. Furthermore, the vision of human progress I have known has seriously faded. It had egregious faults, but it provided a higher plane of argument to which we could ascend as we contested the earthly details of how it might be fulfilled. The visionary void in our collective consciousness has been patched over with increasingly meaningless consumption, free market ideology, and globalization blatantly designed for corporations. Few in academe want to talk about these things, or the roots of fundamentalism or terrorism, internal or external. And yet we may be breaking through this morass as consumption, globalization, and corporatocracy are questioned by grassroots movements and threatened by their own imbalance and dishonesty.

The environmental and energy crises of the 1970s raised serious questions about the role of scientists and progressive governance in democratic societies. For a quarter-century, modern societies have been working out new answers, and considerable progress has been made. At the same time, the complexities of climate change and biodiversity loss, the corporate politics of sound science, and the science wars of academe have introduced whole new issues. For me, environmental provocations and questions of sustainability arise in the interplay of these larger social and scientific challenges that must be addressed but are harder to take on. Yet I know of no better process for coming to grips with these challenges than the transdisciplinary shared learning and scholarly activism that ERG has fostered to grasp what we once thought was an energy crisis.

Acknowledgments

Ned Birdsall, John Harte, and John Holdren helped me keep the details straight. Nancy Rader, Peggy Barlett, and Geoff Chase provided helpful editorial suggestions. A full list of the students who made this program a success and the titles of their research projects and dissertations can be found at <www.berkeley.edu/erg>.

6

Place as the Nexus of a Sustainable Future: A Course for All of Us

Laura B. DeLind and Terry Link

Michigan State University was founded in 1855 as the prototype for sixty-nine land grant institutions in the country today. Enrolling 33,300 undergraduate students, the university buildings and grounds spread over 5,200 acres along the Red Cedar River in East Lansing (population 46,000).

We have come to a place in our history and in the history of our planet where we must acknowledge limits to our physical existence. To survive in any responsible fashion means managing ourselves and our resources in a sustainable manner—leaving equal or greater opportunity for the generations that come after us. *Sustainability* has become a watchword, if not a buzzword, and we are applying much science, technology, and commercial rhetoric to solving the sustainability puzzle.

But if sustainability involves renewed awareness of and interaction with the natural environment, it must also involve equity, economy, social justice, and cultural and spiritual meaning in equal measure. It is less a problem to solve than it is a way of being, a way of relating to a myriad of living creatures and living systems. Such an inclusive and integrated mind-set is conspicuously absent from most institutions of higher learning. To paraphrase Parker Palmer, "We've been great at thinking the world apart; what we need to do now is to think it back together" (Palmer 1998). One such attempt was made at Michigan State University (MSU), spring semester 2001, in the form of a two-credit course: "Our Place on Earth: Experiencing and Expressing Our Relationship to the Natural Environment."

Is it possible for a single semester-long course, taught only once, to have an impact on the university community, its practices, policies, and psyche? Can one course really do much to help us "live sustainably," to "think ourselves back together"? Most of us would say no, but most of us, in this case, would be wrong. The course enabled students to "see" and "feel" the campus, to effect change there and within surrounding natural and built environments. It succeeded in bringing people together across institutional divides (e.g., students, faculty members, operations staff, and administrators) to share expertise and stretch personal assumptions. The course also succeeded in catalyzing new academic and practical efforts designed to give meaning and purpose to a pedagogy of place.

What follows is the story of this course—its background, premise, objectives, design, and outcomes. What we learned as it unfolded and what we continue to learn from the energy it unleashed is offered up in the spirit of discovery. For us, it is a bright and vital spot in a very long journey toward campus sustainability.

Background

In 1997, MSU Green, a grassroots effort of five students, faculty, and staff, called for a campus-wide environmental assessment and the formation of a university committee to advocate for a sustainable campus. Their proposal was well timed. The academic governance system was about to retire a moribund committee responsible, in part, for overseeing the "academic environment." At the same time, a publicly beloved if somewhat seedy greenhouse was slated for demolition. Both "deaths" focused heightened attention on the campus environment.

While the proposal was defeated by a single vote in Academic Council, MSU Green was asked to reshape it, working with academic governance committee members on a consensus basis. The year-long process led to an agreement approved in the fall of 1998: the creation of the University Committee for a Sustainable Campus (UCSC). As part of this agreement, a sunset clause called for the committee's review after three years. In 2002, UCSC withstood its first review and was authorized to continue for another three years.

UCSC began its work in January 1999. An eighteen-member committee comprising students, faculty, operations staff, and administrators created a mission statement, identified goals, and outlined potential projects. Recognizing that speed was of the essence and that little would happen quickly without someone in charge, UCSC members (most notably the present authors) applied for a U.S. Environmental Protection Agency Sustainable Development Challenge Grant to fund an Office of Campus Sustainability and hire a program director. Nearly 1,000 proposals were submitted to the EPA; twenty-seven were funded, UCSC's among them. As one of its promised actions, the proposal called for the development of courses on sustainability open to the entire university community.

To this end, UCSC organized a semester-long seminar series, "Assessing MSU's Environmental Footprint," in spring 1999. The university community was invited to review MSU's water, energy, and solid waste management practices, their environmental impacts, and alternative scenarios. This thirteen-week series, sponsored by over a dozen departments, programs, and colleges, brought "experts," students, and laypersons together and catalyzed a number of working relationships. Most notable was the development of a committee to study the Michigan State watershed and to develop a wellhead protection team. Out of this concern grew MSU-WATER (Watershed Action Through Education and Research), a twenty-two-member committee, representing fourteen departments and off-campus stakeholders, that works in transdisciplinary fashion to safeguard the health of our watershed. Significant funding from both campus and off-campus sources reinforces this work.

In a sense, the seminar series and MSU-WATER presaged how UCSC itself would operate as a committee, for its mission was clearly dedicated to "foster[ing] a collaborative learning culture."[1] Thus, when we began to plan for a second course in spring 2000, UCSC members, and the authors in particular, felt that instead of a purely scientific and technological look at sustainability, we should create a course that bridged the humanities and the social sciences. It was time, we felt, to consider the qualitative as well as the quantitative dimensions of the concept.

This decision was not difficult, as both of us have an abiding interest in the concept of place. Terry Link, a librarian for twenty years with a strong background in environmental studies and geography, has tied the responsibilities of citizenship to the ecological and social fabric of community life. Laura DeLind, an anthropologist carrying out applied research in her own backyard, has used local food and farming as a vehicle for increasing environmental awareness, democratic engagement, and cooperative self-reliance. Since both of us spend half our waking lives on the MSU campus, we were determined to foster an awareness of this place that we share and to use it as a living-learning laboratory.

Our Perspective on Sustainability and Place

Why "place"? How does place relate to the concept of sustainability? For us, the reason and the relationship are seamless. We feel that sustainability, anchored as it is in the earth itself, cannot be taught as a grand abstraction. It cannot be pondered in privileged isolation or treated as a "problem" to be solved through the intercession of "things" and wise market management. Neither, for that matter, can we understand ourselves or our behaviors in disembodied, generic ways. Rather, sustainability—if it is to exist—has to be felt and practiced. People must come to know and care for (show affection and responsibility for) the places they inhabit.

Yet place is a concept of many dimensions—a shape shifter of sorts. It can be tangible, sensual. It can exist under our feet; it can literally ground us, anchor us, give us roots. But place can also be social and spiritual. It can be as intangible as history, as creative as culture, as mystical as creation myths. Instead of something absolute, place can be a matter of shifting identities, shared understandings, and relationships not only among ourselves but among all living creatures.

Place is also particular, unique. There are no two places exactly alike, just as there are no two snowflakes or human beings exactly alike. And this makes it necessary for us to cooperate as well as to make choices. Shall we meet at your place or mine?

But place is also universal. As Steven Jay Gould has written, we—whether as individuals or as a species—belong to, are part of, have a place within the "unforgiving continuity" that is biological life on this planet. "Evolution is 'roots' [our place] writ large" (2001, 48). We are part of—inseparable from—profound, awesome processes that encompass the globe. It is, as Gould explains it, enough to take your breath away.

Place, then, is far larger than contradiction; it is not an either-or proposition. It is at once concrete and abstract, fixed and changing, unique and universal. Place, like life itself, is simultaneously deep and wide. And to start to understand place and our many places, we also must learn to think and act deep and wide. We cannot do only one or the other. We cannot live in the virtual—cannot be everywhere at once—and ignore, disconnect from, the reality outside our doors. Neither can we run away to the wilderness, homestead sixty acres, and let the rest of the world go hang. In both cases, our presence, our weeds, our waste, and our woeful ignorance will take their toll.

If we want to be sustainable, then we need to know our place in every sense of this word. But where are we to begin? Perhaps the best answer we have found is uncommonly simple. We begin with the local, with the embodied, with the personal and the familiar. We begin where we are and where we live. We begin with what we know and what we see before us, and out of this eventually comes the stuff of a greater and deeper wisdom. These, then, were the working assumptions that gave rise to "Our Place on Earth," a course designed to explore the concept of place and MSU as a place (figure 6.1).

Course Objectives

With our perspective firmly defined, we found that we could identify at least four course objectives. First, we wanted to provide multiple opportunities for students to consider their relationship to place and the world around them in new and possibly uncomfortable ways. We wanted to create spaces for what Gary Nabhan calls "discovering." As he explains it, "*Discovering* [is] a process far different from the heroic act of *discovery*. Through the process of *discovering,* we seldom achieve any hard-and-fast truth about the world, its cornucopia of creatures, or its cultural interactions with them. Instead, we are inevitably assured of how little we know about that on which each of our lives depend" (1997, 98). Discovering, then, is a process of being humbled and awestruck simultaneously. It is about finding ourselves enmeshed in a world of great mysteries and about suspending, for awhile at least, our need to judge and control them.

Second, we wanted to provide students with an understanding of the connections that bind them not only to one another but to all places and life forms. Far from being fixed, our bindings and our boundaries are permeable, fluid, forever being reconfigured and negotiated. Our identities are constructed on these living and shifting relationships; so are our responsibilities and the impacts of our actions. What we choose to eat or where we choose to live, for example, is a consequence of existing human and environmental relationships; our choices, in turn, constrain and set in motion forces beyond ourselves. To know ourselves well and to keep ourselves well requires us to understand as fully as possible our

MICHIGAN STATE
UNIVERSITY

Our Place On Earth: Experiencing and Expressing
Our Relationships to the Natural World

Spring Semester Seminar Course
SSC 290, Section 2 (2 Credits)
Thursdays 3:00 p.m. at 107 S. Kedzie

Sponsored By:
University Committee for a Sustainable Campus

Colleges of Social Science, Arts and Letters, Human Ecology, and James Madison; Center for Remote Sensing GIS, Center for Urban Affairs, RISE, Bailey Scholars, Eagle, Sociology, Resource Development, English, Libraries Computing and Technology, Anthropology, Sustainable Agriculture Programs-MSUE, C.S. Mott Chair of Sustainable Agriculture, Women Studies, and Philosophy

Scott Russell Sanders, Jan 11
The Nature of Place
Alan Rudy, Jan 18
Social Construction of Nature
Andrew Light, Jan 25
Urban Places and Nature
Janis Rygwelski, Feb 1
Healing and Nature
Laura Westra, Feb 8
Ecological Integrity: Connecting Environment,Health and Justice
Bunyan Bryant, Feb 15
Our Place on Earth and Environmental Justice
Stephanie Kaza, Feb 22
Practicing in Place: the Body as First Home
Madhu Prakash, March 1
Regenerating Our Places
Michael Shuman, March 15
Community Friendly Business: The Affordable Alternative to Globalization
Dave Dempsey, March 22
Environmental Activism Connecting Passions / Place
Grace Lee Boggs, March 29
Environment, Place and Movement Building
Henrietta Mann, April 5
Nature, Place and Indigenous Knowledge
Charlene Spretnak, April 12
The Integrity of Place
Joan Dye Gussow, April 19
Food, Agriculture, and Resources in Place
Frank Fear & Richard Bawden, April 26
What does this have to do with a land grant university?

For more information contact: Terry Link
E-Mail link@mail.lib.msu.edu
Phone 355-1751

MSU is an affirmative action, equal opportunity institution

Figure 6.1
Flyer for "Our Place on Earth" course, Michigan State University

dependence and our effect on this web of life—if for no other reason than the fact that each of us lives downstream from somewhere (and someone) else.

Third, we wanted to provide students with opportunities to exercise a sense of engagement and empowerment at home. Learning and knowing, we believe, should not be a passive experience. Daily life is not a backdrop to education but education itself. Therein lie the sensibilities and the autonomy necessary for citizenship in a global world. Before losing themselves in the virtual or plunging head long into the international, students need to examine carefully and critically what exists under their feet and outside their front (and back) doors. They need to take their places seriously. As Sanders has put it,

> To become intimate with your home region, to know the territory as well as you can, to understand your life as woven into the local life does not prevent you from recognizing and honoring the diversity of other places, cultures, ways. On the contrary, how can you value other places if you do not have one of your own? If you are not yourself "placed," then you wander the world like a sightseer, a collector of sensations, with no gauge for measuring what you see. Local knowledge is the grounding for global knowledge. Those who care about nothing beyond the confines of their parish are in truth parochial, and are at least mildly dangerous to their parish; on the other hand, those who "have" no parish, those who navigate ceaselessly among postal zones and area codes, those for whom the world is only a smear of highways and bank accounts and stores, are a danger not just to their parish but to the planet. (1993, 114)

Finally, we wanted to create a mobile experience, or put a bit differently, we wanted students to take their place-based affections, sensibilities, and responsibilities with them to new locations. In this way, they, not unlike Johnny Appleseed, would always inhabit real places, bringing with them not only the promise of new life and rooted wisdom but also a sense of coming home.

Course Design

"Our Place on Earth" was a two-credit course, sponsored by a dozen MSU colleges, programs, and departments, and offered under the administrative shelter of the College of Social Science. It met once a week for an hour and forty minutes. Over 100 students registered from schools as diverse as Arts and Letters, Engineering, Veterinary Medicine, Agriculture, and Natural Resources. Within this basic structure, we worked to realize our rather nontraditional objectives through a collage of intellectual, physical, and sensual experience.

We invited fifteen scholars, practitioners, and activists from around the country, as well as MSU, to present their views of place and its relationship to nature and the environment. They spoke about human rights and food production, about entrepreneurship and commonwealth, about culture and indigenous wisdom, about soil and bodies and medicine. As individuals, our speakers were as diverse as their academic backgrounds and subject matter. They were men and women, artists and academics, young professionals and octogenarians, rural residents and urban dwellers, Native American, African American, Asian American, Buddhist, Muslim, and Christian. And for all their diversity, they were united in their profound concern for the earth and in their continued activism on behalf of social justice, democratic process, and the awesome mystery and responsibility of life itself. The collective power of their message was palpable.

Where the language and personal style of one or two instructors might have grown predictable and forgettable, the continually changing language and identity of our speakers was stimulating, if not provocative. With hardly a note card in sight, Scott Russell Sanders spoke of his connections to his home place in Bloomington, Indiana, told stories and read from his lyrical writings. Stephanie Kaza, a practicing Buddhist, used a bell to help focus our attention as she led students in a simple meditation. Mahdu Prakash, a woman of unusual grace and beauty, stunned students by openly talking about "shit" and our literal and figurative need to make soil. Henrietta Mann affectionately embraced the class, calling all of us her grandchildren and speaking to us in her own native language. Finally, with simple stage props, dimmed lights, and admission tickets, Frank Fear and Richard Bawden performed, à la *My Dinner with André,* a critique of the role of the land grant university in today's society. The array of personal artistry made the message far richer than any one or two of us could have managed, or imagined, alone.

Typically, each session began with a public hour-long presentation to which the entire university and Lansing communities were invited. Attendance varied from 100 to 150. The presentations were followed by lively and often lengthy discussions among speakers and students. Students frequently stayed after to talk with guest speakers and occasionally went out to dinner with them.

Each speaker recommended readings that were assembled into a course pack (see the Appendix). Students were required to read these materials prior to each weekly session. Two short written assignments asked them to compare, contrast, and creatively apply the perspectives of several speakers and authors. In all but a couple of cases, the sessions were recorded and the audiotapes given to the MSU Vincent Voice Library.

In addition to the classroom experience, the course required students to attend two tours of the MSU campus (one indoors and one outdoors). Their purpose was threefold: (1) to expose students to areas and operations within the university that they had not "seen" before, (2) to encourage students to consider the relationships these areas and operations maintain on and off campus, and (3) to reflect on the sustainability of these relationships. Students in groups ranging from three to fifteen persons visited the cyclotron, the laundry, the bakery, the research farms, the power plant, the observatory, and university salvage. They took tours of campus wood lots and teaching gardens; they canoed down the Red Cedar River.

A final assignment asked students to reflect on their tour experiences and to relate them to course presentations and readings, as well as to their own developing sense of sustainability and the campus as a place. Students saw connections between the outer space of astronomy and the inner space of nuclear physics. They questioned the ethics of high-tech agriculture and the absence of local food in the dining halls. They found parallels between the loss of natural areas on campus and the loss of local businesses on Main Street. This experiential component was exceptionally well received. Not only did students find the tours eye-opening, but they recommended that a separate course be created to introduce all students to the physical reality of the MSU campus.

Course Evaluation

The written evaluations we received from students and visitors were overwhelmingly positive. They committed to paper what we frequently heard in the classroom: that people want to connect and to belong in real, multidimensional, and lasting ways to something greater than themselves. They want to know more fully how they relate to the world about them. They want to reinhabit their places. One extremely bright, articulate, and focused prelaw senior, by contrast, resented (and frequently derided) the "nonrational," expressive nature of many of the course speakers and readings. Yet even his resistance wavered a bit after a canoe ride down the Red Cedar River with a naturalist. He allowed, in a final parting note to the instructors, that "this was a very intriguing course with some very different ways of looking at the world." More typical, however, were comments like the following:

"I can honestly say that this two credit class that meets only once a week has taught me more than any other class I've taken at MSU. The ideas, thoughts, reflections from the excellent selection of speakers will stay with me for a lifetime."

"This course has caused a reawakening of self and intellect. I feel I can work to be a better 'whole' in my community. I am more aware!"

"This class was more 'deep' than a math or ATL class (by far!).[2] If all my classes were as interesting as this one, my grade-point would skyrocket."

"The topics raised provoked many thoughts and produced some life style changes. I've begun to regard this place as holy, become an even more conscientious consumer and have really stood up for who I am and values I have assimilated through this class."

"The SSC 290 lectures and discussions were superb! The people you brought in and the topics presented were not only very stimulating but they certainly pointed to new and enlightened ways to view our planet and our place on this planet. Walking back to the Natural Resources building from Kedzie I noticed that even my gait was different—*lighter*. Your speakers were a source of inspiration to me each week I was able to attend. To me it was one of those uplifting experiences that remind me how fortunate we are to work in a university environment."

Students also made suggestions through oral and written evaluations about how we might fine-tune the course should we offer it again, an option they heartily endorsed. They suggested expanding it from two to three credits and providing separate and smaller discussion sessions. The absence of regular class time devoted to discussion in advance of or directly following each guest speaker was keenly missed.

As with most other things, time proved to be our greatest constraint. Originally we talked about including an "expressive" component. We wanted students, as audience or as participants, to engage in art performances—dance, exhibitions, concerts, storytelling—that spoke in some sensory way about place. We did not get around to it. We also realized midway through the course that we should have planned for a service-learning component. This, we feel, would have offered students new ways to connect with the university and greater Lansing community and enabled yet another experiential, possibly activist, perspective. Such opportunities for a living pedagogy may be possible within an expanded three- or four-credit course format.

It was apparent that the course affected bodies as well as minds and inspired personal and social action. Three young women impressed by the environmental and health advantages of local food joined a working CSA (community-supported agriculture). A student who took the canoe trip launched a petition, ultimately signed by 100 other MSU students and delivered to the head of Campus Parks and Planning, to create more courses that directly integrate the living campus into the classroom. Several other students became involved in a campaign to save a local bagel shop from competing national franchises. When this failed, they began to work on identifying, protecting, and promoting locally owned businesses in East Lansing.

Course Outcome

"Our Place on Earth" was designed to blur traditional intellectual and physical boundaries and to challenge the familiar. We believe it succeeded. It introduced over 100 students from the nonprofessional colleges at MSU and 30 to 40 community members to numerous, alternative ways of knowing that were humanistic, qualitative in nature, and open ended. It asked them to be self-reflexive and to rethink the authority of science, the virtue of the virtual, and the prevailing forms and purpose of pedagogy. It lent legitimacy to experiential learning and the particular. It encouraged students to apply both theory and action in equal measure and to readmit the sensual, the social, and the sacred into the realm of the academic.

Just as we had help bringing the course to life, we recognize that many people in addition to ourselves continue to use it as a springboard for place-based learning and discovery. It remains alive in the minds and affects the behavior of many people. It has moved well beyond its scheduled semester-long existence.

One example has been a series of three UCSC-sponsored natural area tours to which the university president, provost, vice presidents, and deans were expressly invited. The purpose of these tours was to bring administrators face to face with the "outside"—with the awesome beauty and complexity of the MSU campus—and to catalyze a new position, possibly the first at any land grant institution: that of—campus naturalist. One vice president was so enthralled by the tour that he brought his spouse back the following weekend for a walk through the wood lot. The president said that he would bring his grandchildren through soon. At this time, however, budget cuts have eliminated the possibility of creating a campus naturalist position, though we are hopeful that outside funding can be found.

In spring 2002, UCSC offered a third course, "Earth Charter: Pathway to a Sustainable Future?" Using the United Nations Earth Charter as a vehicle to discuss the range of issues involved with sustainability, this course responded to earlier student evaluations. We made it a three-credit course with a separate discussion section. We included a service-learning component and allowed for multiple approaches to demonstrate learning. We also used predominantly local guest lecturers. Of additional interest, six students from "Our Place on Earth" also enrolled in this course.

Inspired by "Our Place on Earth," the director of student affairs in the College of Social Science asked Laura to help plan and teach a freshman seminar for fall 2002 called "Getting to Know (This) Place." It is designed to provide students who are quite literally "between home places" with the tools for thinking about and exploring the MSU campus. Tours and performances, poetry readings and discussions, indoors and out, will help orient students intellectually, emotionally, and physically to the university and the surrounding communities of which they are now a part.

"Our Place on Earth" also catalyzed a group of eight students and faculty members (half of whom had attended the course) to write a concept paper that was submitted to the W. K. Kellogg Foundation to expand the university's understanding of and engagement with local food and food systems. Although the proposal was not funded, UCSC did create the Sub-committee on Campus Food and Agriculture. It now includes the manager of University Food Stores and the coordinator of Dining Hall

Services and designed two "sustainable food events" (local and organic meals) for the 2002–2003 academic year.

There are some individual stories to tell here as well. John, a retired state natural resources specialist, offered to give canoe tours of the river for the course. When this did not prove to be sufficient involvement for him, he came to a number of the presentations and often brought his wife, an upper-level university administrator, with him. As a result, and in association with several students inspired by his tours, he has revived the Friends of the Red Cedar. This group, now chaired by a student who was introduced to the river through "Our Place on Earth," has hosted two river clean-ups since the course began. John's love affair with the Red Cedar and its revival found a host and a vector.

Diane is a Ph.D. candidate and an adviser to a special undergraduate program in the College of Agriculture and Natural Resources. She brought students from her program to sit in on "Our Place on Earth" lectures and provided a reception for one of the speakers. She has subsequently been involved with UCSC in writing grant proposals, working on curriculum infusion efforts, and attending workshops with others from the university on campus sustainability.

One final story will illustrate just how circuitous and endlessly emergent the connections triggered by "Our Place on Earth" can be. A full year after the course ended, Laura was thinning carrots across from a new member of her community farm. As they talked, the new member said that she was a professor of education at MSU. She had attended one of the course lectures and had learned a bit about the work of UCSC. Would it be possible, she asked, to get involved? She is now the newest member of UCSC's Sub-committee on Campus Food and Agriculture.

Reflections

"Our Place on Earth" buoyed our spirits. It gave legitimacy to ideas and ways of being that are not easily codified and are too easily marginalized. It allowed us all—organizers, lecturers, tour guides, students, and visitors—to grapple with personal uncertainty and public dis-ease in ways that strengthened our respect for community and our ability—in Gary Snyder's words, "to lay claim to the term *native* and the songs and dances, the beads and feathers, and the profound responsibilities that go with it" (1995, 236).

Becoming native to MSU we understand is a forever process. "Our Place on Earth" moved us gently in that direction. It introduced us to great and previously unknown thinkers in philosophy, movement building, democracy, racism, and simplicity. It encouraged us to collaborate on projects with persons we did not know but who felt that the process of collaborating was as valuable as any "product" created. We found, too, that the often messy and meandering process of building interpersonal trust was the key to sustainable change. Finally, the course reaffirmed our ability to teach ourselves and to find and reinhabit our places, whether at a land grant university, on an urban street corner, or in a community garden.

Readings for Our Place on Earth

Scott Russell Sanders January 11

Scott Russell Sanders. "Local Matters" in Scott Russell Sanders, *Secrets of the Universe*. Boston: Beacon Press 1991. pp. 96–103.

——— "Hunting for What Endures" in Scott Russell Sanders, *In Limestone Country*, Boston: Beacon Press, 1991. pp. 1–6.

——— "Landscape and Imagination" in Scott Russell Sanders, *Secrets of the Universe*. Boston: Beacon Press, 1991, pp. 83–95.

——— Staying Put. Boston: Beacon Press, 1993, pp. 97–121.

Alan Rudy January 18

Donald Worster, "The Ecology of Order and Chaos." *Environmental History Review* 14 (1–2): 1–18 (1990)

Andrew Light January 25

Avner de-Shalit, "Ruralism or Environmentalism?" *Environmental Values* Vols. 5, No. 1, 1996, pp. 47–58.

Bill E. Lawson, "Living for the City: Urban United States and Environmental Justice," in *Faces of Environmental Racism* eds. Laura Westra and Peter Wenz Lanham, Md. Rowman and Littlefield Publishers, 1995), pp. 41–55.

Andrew Light, "Elegy for a Garden: Thoughts on an Urban Environmental Ethic" *Philosophical Writings*, forthcoming, Vol. 14, 2001.

Janis Rygwelski February 1

Riley, David S. "The Mystery of Health: Reclaiming Medicine's Soul" *Alternative Therapies* vol. 3 March 1997 pp. 128–127.

Cumes, David. "Nature as Medicine: The Healing Power of the Wilderness" *Alternative Therapies* vol. 4 March 1998, pp. 79–86.

Laura Westra February 8

Pimentel, David, Westra, Laura and Noss, Reed. *Ecological Integrity: Connecting Environment Conservation and Health*. Washington, DC: Island Press. 2000. Chapter by Westra.

Westra, Laura. "Environmental Risk, Rights and the Failure of Liberal Democracy: Some Possible Remedies", in Laura Westra *Living in Integrity: A Global Ethic to Restore a Fragmented Earth*.

Lanham, Md.: Rowman & Littlefield, c1998. pp 53–80.

——— "The Faces of Environmental Racism: Titusville, Alabama and BFI" in Laura Westra and Peter S. Wenz. *Faces of environmental racism: confronting issues of global justice*/Lanham, Md.: Rowman & Littlefield Publishers, c1995. pp. 113–133.

Bunyan Bryant February 15

Bryant, Bunyan and John Callewaert. "Why is Understanding Urban Ecosystems Important to People Concerned About Environmental Justice". Draft paper by authors. Not published.

Stephanie Kaza February 22

Kaza, Stephanie. "Acting with Compassion", *Ecofeminism and the Sacred,* ed. Carol Adams, Continuum, 1993, pp. 50–69.

Barnhill, David. "Great Earth Sangha: Gary Snyder's View of Nature as Community, in *Buddhism and Ecology,* ed. ME Tucker and DR Williams, Harvard University Press, 1997, pp 187–217.

Snyder, Gary "The Porous World", *A Place in Space,* Washington DC: Counterpoint Press, 1995, pp 192–198.

Kaza, Stephanie. "The Attentive Heart", *The Attentive Heart,* NY: Ballantine, 1993, pp 157–165.

Madhu Prakash March 1

Prakash, Madhu Siri. "From Global to Local" in Gustavo Esteva and Madhu Suri Prakash. *Grassroots Post-modernism: Remaking the Soil of Cultures*. London; New York: Zed Books; New York: Distributed in the USA exclusively by St. Martin's Press, 1998.

——— "Beyond the Individual Self" in Gustavo Esteva and Madhu Suri Prakash. *Grassroots Post-modernism: Remaking the Soil of Cultures*. London; New York: Zed Books; New York: Distributed in the USA exclusively by St. Martin's Press, 1998.

——— "Grassroots Postmodernism: Refusenik Cultures" in Prakash, Madhu Suri, and Gustavo Esteva. *Escaping Education: Living as Learning within Grassroots Cultures*. New York: P. Lang, c1998.

Michael Shuman March 15

Shuman, Michael H. "Amazing Shrinking Machines," *New Village* no.2 2000, pp. 17–32

——— "Community Entrepreneurship" *Shelterforce* September/October 1999, pp. 10–13.

——— *Going Local: Creating Self-reliant Communities in a Global Age.* New York: Free Press, 1998. Chapters 2–3.

Dave Dempsey March 22

DeBlieu, Jan. "Mapping the Sacred Places" *Orion* Spring 1994, pp. 18–24.

Grace Lee Boggs March 29

Boggs, Grace Lee. "One Thing Leads to Another: Cooperative Developments in Urban Communities." Keynote Address, Michigan Alliance of Cooperatives, E. Lansing, Michigan, October 20, 2000. *http://www.boggscenter.org/co-onethg.htm*

Henrietta Mann April 5

Cajete, Greg. "An Enchanted Land: Spiritual Ecology and a Theology of Place" *Winds of Change* Spring 1993, pp. 50–53.

Whiteman, Henrietta. "White Buffalo Woman". In

Charlene Spretnak April 12

Spretnak, Charlene. *States of Grace: The Recovery of Meaning in the Postmodern Age.* San Francisco:Harper, 1991. p. 112–113

———. *The Resurgence of the Real: Body, Nature, and Place in a Hypermodern World.* Reading, MA: Addison Wesley, 1997. pp. 122–123 (plus footnotes 55 & 56 on p. 240)

Chapter Five, "Embracing the Real" (pp. 181–215 plus footnotes on p. 243)

Joan Dye Gussow April 19

Gussow, Joan "Can a Community have a Food System?" *Open Spaces* 2:2:12–13, Spring, 1999.

Schwartz, David. "A gift of a garden." *Smithsonian* September, 1997, pp. 67–71.

Pollan, Michael. "Playing God in the garden." *New York Times Magazine* October 25, 1998, pp. 44–52.

Ho, Mae-Wan. "One bird—ten thousand treasures," *The Ecologist,* October 1999.

Kloppenberg Jr., Jack, Hendrickson, John and Stevenson, G.W. "Coming in to the Foodshed." *Agriculture and Human Values* 13(3) Summer 1996 pp. 33–42.

Frank Fear and Richard Bawden April 26

Campbell, John R. *Reclaiming a Lost Heritage: Land-Grant and Other Higher Education Initiatives for the Twenty-first Century*. Ames, IA: Iowa State University Press, 1995. Chapter 1, pp. 3–27.

Orr, David W. *Earth in Mind: On Education, Environment, and the Human Prospect*. Washington: Island Press, 1994. Chapters 1–3, pp. 7–34.

Notes

1. "The mission of the University Committee for a Sustainable Campus is to foster a collaborative learning culture that will lead the MSU community to a heightened awareness of its environmental impact; to conserve natural resources for future generations; and to establish MSU as working for creating a sustainable community." The UCSC web site is <www.ecofoot.msu.edu>.

2. ATL (American Thought and Language) courses are the basic writing courses that all students must take, generally during their freshman or sophomore year.

References

Gould, Stephen Jay. 2001. "I Have Landed." *Natural History* 109:46–59.

Nabhan, Gary Paul. 1997. *Cultures of Habitat: On Nature, Culture and Story.* Washington, D.C.: Counterpoint.

Parker, Palmer. 1998. *The Courage to Teach: Exploring the Inner Landscape of a Teacher's Life*. San Francisco: Jossey-Bass.

Sanders, Scott Russell. 1993. *Staying Put*. Boston: Beacon Press.

Snyder, Gary. 1995. "The Rediscovery of Turtle Island" in *A Place in Space: Ethics, Aesthetics, and Watersheds: New and Selected Prose*. Edited by Gary Snyder. Washington, D.C.: Counterpoint.

7

Building Political Acceptance for Sustainability: Degree Requirements for All Graduates

Debra Rowe

Oakland Community College (OCC) is one of the largest community colleges in the United States, with an enrollment of over 24,000 students and 280 full-time faculty. The faculty have a powerful union and a strong tradition of academic freedom. OCC is a four-campus structure spread throughout a very affluent county northwest of Detroit with areas of poverty within and just outside its boundaries. OCC has the largest first-year class of any higher education institution in the state of Michigan.

Global environmental awareness, social responsibility, and interpersonal skills are required components of all degrees at Oakland Community College (OCC). It has been a large task to bring about these requirements and took twelve years to accomplish. Although OCC has a national reputation as an innovative college, creating change in such a large institution has been a difficult challenge. It is my hope that understanding our success in building sustainability-based graduation requirements at OCC will be useful precedents for other educational institutions.

In my early years of teaching, I dismissed the possibility of such a requirement as politically impossible, given the power structure, the territoriality of the faculty, and the dominant philosophical-conceptual paradigm. Yet I was moved to action by the fact that almost all our students completed their degrees without any course work in environmental literacy, much less sustainability as a larger framework. This narrative reviews the strategies that were helpful in building a critical mass of support: mentoring, reframing, futuring, environmental scanning, nurturing project champions, creating discussions to rise above territoriality, sculpting presentation of concepts, counting the votes, influencing others' opinions, handling difficult people, creating politically acceptable choices, building powerful coalitions, implementing key strategies, and coping effectively with the personal cost of the process.

Background: Environmental Awareness in the Community

After owning a renewable energy and energy audit company, I chose in 1980 to become a full-time faculty member in OCC's Technology Department. I was hired to create and teach courses in renewable energies and energy efficiency. However, my goal extended beyond the institution; I wanted to improve the environmental literacy of metropolitan Detroit. First, I started an Energy Awareness Center out of my classroom and offered open houses and conferences for the public. My outreach efforts created informal partnerships with the state energy office and environmental and professional associations to provide targeted training programs and conferences, and build green networks for architects, engineers, facility managers, builders, and consumers. The model equipment purchased for the courses provided multiple examples of environmental sustainability for the community, including energy-efficiency technology, solar heating systems, aquaculture, solar greenhouse food production, permaculture, and a 4 kilowatt solar electricity research and demonstration project. The college supported my activities, as they also met the need to market the alternate energies technology courses and related degrees. At the same time, I built a relationship of friendship and trust with the college media person and continually created news stories to get media coverage on the benefits of renewables and energy efficiency.

As the interest in renewable energies waned in the 1980s, I went back to graduate school for psychology and business degrees to ensure my tenure at the college and therefore the continued offering of alternate energies technology courses. I now teach psychology and alternate energies courses at OCC, both from a sustainability perspective. I am still committed to the larger community and currently use U.S. Department of Energy funds to help other colleges around the country establish energy efficiency, solar and wind energy, and other renewable energies courses and degrees. (See <www.ateec.org/pete> for details and the Sustainability Online Handbook for K–12 teachers at <www.urbanoptions.org/SustainEdHandbook>.)

Environmental Literacy for All Graduates

After spending ten years educating the public and professionals, I turned my attentions inward to the college. I knew the vast majority of the

24,000 students at OCC would never take alternate energies technology courses or environmental science courses and that as an institution, no efforts were being made to ensure that all our graduates were environmentally literate or aware of the sustainability paradigm. I realized that grappling with campus politics was the next step to try to ensure environmentally literate, sustainability-oriented graduates.

The task was somewhat daunting. The concept of sustainability was not well known and was not going to be easily accepted within the college. Most of the administrators and faculty had received their formal education when the prevailing paradigm was "man conquering nature" instead of the sustainability paradigm. I believed that politics and personalities would affect, and possibly hinder, any efforts to add degree requirements about sustainability. I looked for possible coalitions and strategies to enhance the chances for political acceptance and institutionalization of sustainability curricula.

A Multipronged Approach

The process for establishing sustainability degree requirements was multifaceted, for two reasons. In case I met a dead end, other avenues for change would still be active. Second, sharing the vision in multiple forums and formats is a necessary step to transform the organizational climate and the perceived norms of the college. I knew from my Ph.D. in business that to change norms, people need to see the vision many times and in multiple ways so they can cogitate on it, understand it, and internalize it. I had to remind myself constantly to be patient and communicate clearly. As Richard Thompson, the OCC chancellor, said, "Activists need to pull back from the object of their zeal, and look at the process by which it can be realized." I also knew the strategies I used had to include the building of mutual respect and caring, not only because they are part of the sustainability paradigm but also because I am personally committed to these values in my interactions with others.

The new product diffusion of innovation curve shows that there are different types and sizes of groups when it comes to the adoption of new concepts: approximately 2.5 percent are innovators, 13.4 percent are early adopters, 34 percent are the early majority, 34 percent are the late majority, and 10 percent are the laggards (Kinnear and Bernhardt 1986). I knew that sustainability concepts had to get support from the innovators, the early adopters, and enough of the early majority to generate a critical mass. Given my longevity with the institution, I knew who the innovators and early adopters at the college were. I had worked with them on other projects, or they had expressed their support for my early efforts. It seemed to me from previous projects at the college that once we had the innovators and early adopters and we had published the vision and expectations multiple times in well-respected venues, the early majority would tend to go along because of a desire to follow the rules and conform. Laggards are not a problem unless they are active resisters. Therefore, I needed specific strategies to address the concerns of those few active resisters who could obstruct the project. I knew who the active resisters were because throughout the process, I asked many colleagues how the ideas were being received so I could gather the names and the issues of the active resisters.

Gaining Access to Funds and Authority through Mentoring and Reframing

The initiative to create sustainability-based graduation requirements needed funding and support from people in high places. Developing connections with mentors and using reframing were two early strategies that provided ongoing access and helped build strong relationships to those in power. These strategies helped produce support at various levels during key stages of the initiative.

I learned the benefits of being mentored and thereby building relationships with the highest authorities almost by chance. In the early 1980s, the vice chancellor of the college had sent out information about a curriculum in human ecology and asked who might be interested in working on such a curriculum at OCC. I responded out of my natural interest, met with him regularly for a while, and found him to be a mentor to me in many ways. I told him so after one of our meetings, and it had a powerful effect, increasing our communication and our trust in one another. I now consciously take the same approach with other authority figures. To establish these relationships, I honestly identify what I admire in the person's behavior, share my observations, and explain that I see the person as a mentor in this area. After building the relationship, I describe my initiative using the reframing I describe next and ask for ideas about how to accomplish it. Later, if I encounter other administrators who I think may not support the initiative out of a political fear of the risks of change, I mention I talked with their boss, who shared ideas about how to achieve it. Often, this lowers their political concerns, they become more open to learning more about the concepts, and they are often more supportive, or at least act neutral. I have developed some wonderful mentors this way, have increased my skill base by learning from them, have received increased support for many initiatives, and have made many friends as well.

Reframing is metaphorically putting a new picture frame around an idea, changing the verbiage to match the language and include the interests of those in authority. In 1985, a new chancellor arrived at the college. The chancellor was not very interested or educated in sustainability, but I found out we were both members of the World Future Society. Through reframing, I approached the chancellor about starting a Futures Institute, offering futuristic activities within the college and the surrounding community. He loved the idea and funded it. I then created activities for the college and the public with a sustainability focus, since an environmentally sound and more socially just world are essential pieces of a positive future society. By reframing, I could move the sustainability agenda forward and share the sustainability concepts with the chancellor simultaneously.

Environmental Scanning

In 1990, the Futures Institute instituted environmental scanning as a way to educate the college community about the need for change toward a sustainability paradigm. Environmental scanning is a systematic collection of external information in order to lessen the randomness of information flowing into the organization and provide early warnings for managers of changing external conditions (Aguilar 1967). James Morrison, professor of education at the University of North Carolina at Chapel Hill, applied the environmental scanning process to the higher education setting, and we used his work to design our own process (Morrison 2000). We assessed trends in key areas of society that might affect the future of education generally and of OCC specifically.

The college's administration adopted environmental scanning as the first step in their four-step strategic planning process. As director of the

Futures Institute, I became cochair of Environmental Scanning and helped build the membership of the scanning committees to include the four campus presidents and over sixty faculty and administrators. Some faculty volunteered immediately; many joined because we called them with a personal invitation, telling them that their expertise would make a valuable contribution.

We had six committees, and each committee scanned literature about one of the following topics: Educational Trends, Occupational Trends, Economic and Funding Trends, Political/Regulation/Legislation Trends, External Opinion, and Enrollment Trends. Committee members wrote summaries of relevant articles, which were summarized into a committee report and eventually combined into an executive summary that identified ten key trends to be addressed by the college community. I chose to chair the Educational Trends and Occupational Trends committees. While conducting the scan, I hoped the literature review would support the components of sustainability-based education—the need for students to be able to articulate social justice issues and environmental problems that are diminishing the quality of life and the habitability of the planet, describe possible solutions, and develop the social responsibility, citizenship, and change agent skills to help solve these problems. I knew that without the social responsibility, citizenship, and change agent skills, an understanding of societal problems had the potential to increase student apathy (Rowe 2000).

Publications concerning the need for educated citizens to address global environmental problems such as air pollution, ozone depletion, and greenhouse gases emerged in our scan. Our literature search also provided many researchers describing the need to reduce student apathy about societal problems and help students build an increased commitment to social responsibility. The scan in addition produced many documents describing why students need better intrapersonal and interpersonal skills. In fact, this last one was a very high priority for employers, who claimed they wanted employees with better emotional intelligence, conflict resolution, and teamwork skills. Part of my job was to distribute abstracts and articles to the scanning volunteers. I tried to make sure that articulate, assertive, and politically neutral faculty were assigned to read and summarize key articles supporting sustainability education.

Identifying and Nurturing Project Champions

By listening to my colleagues, I found potential project champions during the environmental scan. I nurtured their growth through phone calls, informal meetings, and again building what often became rewarding friendships. In truth, I do not think any of them had the passion for sustainability that I did, though many of them had their own principles regarding quality education, which made them open to the ideals embedded within the sustainability paradigm. Many were passionate about other issues, and the sustainability trends were just a part of the package we worked on together. Nevertheless, out of the environmental scan came a small group of faculty who were committed to implementing a powerful new vision for the OCC curriculum.

Building Support for the Need for Curricular Change

After nine months in 1991, the executive summary of the first scan was presented to the upper administration and the College Academic Senate, and copies of the report were sent to all faculty and administrators. The college structure includes a Campus Academic Senate at each of the four campuses, which together form the combined College Academic Senate. The College Academic Senate has a very strong influence on academic policies throughout the college, and though the administration's strategic planning process had produced a clear document, it now had to be adopted by the Academic Senate.

People often resist change because of their comfort level with the status quo, as well as fear that change will produce loss of identity, control, meaning, or belonging. Human attitudes that support change include a desire to maintain control over the external environment, a feeling that change is both possible and good, a belief that doing and changing are better than accepting situations, and a conviction that the future will be better than the past (Adler and Jelinek 1986).

Therefore, as part of the environmental scan presentation to the College Academic Senate, we showed how status quo curricula would cause loss of control for the faculty. We described how a few state legislatures had taken control of the graduation and other curricular requirements when legislators thought colleges or universities were not responding to societal trends and needs. We discussed the importance of making changes to build support for the college's millage. We framed the information by saying that changing was the best way to maintain control and that the future for faculty and students would be better if we responded to these trends. We moved to action by closing the presentation with an audience brainstorm of how we might begin to make changes college-wide to address these trends.

The College Academic Senate created the Curriculum Research Task Force to address these issues. Of course, this did not happen by chance! Suggestions to form a task force were made to its leadership from many individuals who were vested in seeing action from the results of the scan, including members of the core scanning committees.

As we moved along in this process of building momentum for change, we nurtured the support of upper administration and at least one project champion on each campus, as well as the College Academic Senate leadership. Their support grew as we included them in discussions of the vision, asked them for their opinions regarding our accomplishments to date, thanked them for their past support, and clearly explained what we needed from them. We also continually took the temperature of the informal opinion leaders in the college community regarding the initiative and responded accordingly to try to maintain a positive atmosphere.

Reducing the Negative Effects of Territoriality

The Curriculum Research Task Force chose not to adopt the scan summary, but to recommend the senate study the trends themselves. Yet another committee was formed to do a literature review, which was discouragingly slow but had the result of increased awareness and political buy-in of more faculty.

This committee asked to be put on the agendas of the four campus senates so the faculty and academic personnel could brainstorm answers to the following question: What do our students need from college to be successful in their adult roles of family member, worker, and educated citizen? Using this question was a crucial strategy, since it produced a discussion that rose above the territoriality of the departments and the disciplines. The task force combined the literature review summary with the campus discussion results to produce a list of core competencies that graduates should have for adult success. Taking the time for brainstorming sessions at each of the campuses built support for this list of core competencies.

Some of these competencies were very common in higher education, such as critical thinking, scientific and technological literacy, and communication skills. Three of these competencies were the three sustainability-related items that had been identified initially in the environmental scan: awareness of the global environment, social responsibility, and interpersonal and intrapersonal skills. Combining the sustainability-related competencies with very common, politically acceptable competencies into one list helped get the needed votes for the overhaul of the general education curriculum. In 1999, the College Academic Senate passed the competencies and established a committee to develop a plan to implement them in the curriculum for all undergraduates (see the box).

Sculpting, Swaying, and Counting the Votes

As I helped to facilitate the discussions about the key question, I watched to make sure the sustainability items were included. I even asked colleagues prior to the meeting to raise these issues. I tried to sculpt their presentation to make sure they were included effectively in the discussions. Often, other faculty put these ideas forth anyway, but it's good, before any meeting begins, to have a backup plan.

Even with these strategies, we tried to leave nothing to chance. Faculty supporting the competencies attended all of the meetings or talked to the voting members of existing committees in enough numbers to help ensure passage of the changes. The voting members of the Academic Senate are a finite list, so it was possible to have targeted conversations with members and to anticipate vote counts. Counting votes ahead of time and, if possible, not taking the vote unless victory is clear, is an important political strategy. We used this strategy repeatedly to push forward the competencies agenda.

Winning votes is often about relationships. People were often persuaded during our targeted conversations with them simply by sharing information and answering their questions. Although I always preferred to build understanding of the need for the competencies, we learned that

The General Education Attributes at Oakland Community College

1. Communicate effectively
2. Think critically and creatively
3. Solve problems analytically, systematically, and insightfully
4. Develop an aesthetic awareness
5. Acquire interpersonal and personal development skills
6. Learn independently and collaboratively
7. Be technologically and scientifically literate
8. Appreciate diversity and commonality
9. Develop a strong commitment to social responsibility
10. Understand the global environment

we had to recognize the importance of other dynamics as well. For example, there was often a tit-for-tat dynamic in effect. If we provided support for something we could agree with that they cared about, they would provide reciprocal support for our ideas. Sometimes people seemed to cast their votes because they were flattered by the personal attention and the content seemed secondary, or because they wanted to support a friend involved with the competencies. Within the meetings, we learned it was important to use well-placed comments. Our experience suggests a minimum of three articulate supporters in the room, not sitting next to each other, who speak to the issue either consecutively or close to it in order to have a good chance of influencing the undecided votes in the room. If we let the ideas become too identified with one person, the discussions were not as fruitful. From my business school days, I found Cialdini's book (1984) on persuasion an interesting read. By reading it, I was better able to prevent persuasion techniques being used against our issues.

We learned never to throw away documentation. Many times, we saw college staff get involved midway or later into the process, and they often questioned the validity of past actions. For example, at OCC (as well as at three other institutions I have researched), faculty who got involved later questioned whether the environmental attribute is about the physical environment or only the cultural and business global environment. It was important to have the minutes and research summaries to show them that the physical environment was a key component.

Handling Difficult People and Coping with the Process

In building our critical mass, we dealt with some people who attempted to obstruct the process (without sound arguments, in our opinion). We found ourselves spending a lot of meeting time on the repeated interruptions from people with personalities who were attached to getting attention, to being cynics and naysayers, and to showing their power. They even made public personal attacks on people who dared to disagree with them, and they showed no respect for the work that was completed and the emerging consensus. Such meeting dynamics can be very draining to deal with, so we developed a set of coping skills.

We reminded ourselves not to take attacks personally. Even if it feels personal, angry criticism derives from the speaker's internal landscape, and it is easier to address the critic by avoiding a hurt, victim stance. If I had a strong emotional reaction to a behavior, I learned to avoid being reactive and to use the emotional reaction strategically instead. Once, when a person kept cutting me off and being hypercritical in a small subcommittee, I just gave an exasperated sigh and said, "I have to leave for a minute." This helped the speaker realize she was being inappropriately rude and changed her behavior to a more professional demeanor.

Assertiveness techniques are a set of tools I use when being interrupted or challenged and are available in a number of self-help psychology books. For example, I have found the broken record technique useful. Calmly repeating, "May I finish my sentence?" highlights to others in the room the behavior of the difficult personalities and tends to discount their arguments.

Another helpful approach is to pause, speak in a slower and quieter voice, thank them for their interest, provide clear and kindly worded answers, and then change the topic by posing another question for another member of the committee. The difficult personality may also be a potential learner who may need extra explanations in private. With one difficult personality, a group of us successfully passed a code of conduct at the senate level, so when the person was rude and attacking, we could gently ask him or her to keep comments within the code of conduct. This reminder had a powerful dampening effect on inappropriate behavior.

As an activist, it is important to love yourself as passionately as you do your commitment to sustainability. I am committed to empowering and nurturing others but I used to forget to nurture myself. Although I had some support from colleagues and friends, this initiative took a total of twelve years to date, and many of the involved faculty moved on to other projects. At times, I felt alone as I carried the commitment. I reminded myself that without self-nurturing, burnout would reduce my effectiveness. I collected a bag full of tools to prevent burnout. For example, I consciously revitalized myself by creating playful moments, breathing deeply and enjoying all my senses, and celebrating all of the baby steps of success. I reached out to supportive and encouraging friends and colleagues. I took actions to foster my physical and mental health and laughed good-naturedly as often as possible.

The Ups and Downs of Implementation after the Vote

After the passage of the core competencies, we tried asking the faculty to fill out forms to describe their core competency learning activities and how they planned to assess them at the end of each course. When we encountered strong resistance to the assessment component of our efforts, we had to back off from this strategy to encourage core competency implementation. It taught me a valuable lesson: there is room to make mistakes, change directions, and still keep an initiative going.

To move forward, we needed a new, politically viable way to institutionalize the competencies, and the general education requirements provided that focus. In 1995, the College Academic Senate created an ad hoc committee to review the general education core requirements for all degrees. The committee renamed the competencies "attributes," in order to distance them from some of the ill will caused by the assessment matrices and added two more attributes to the list.

Two more years of literature review, senate discussions, meetings, and votes produced a revised philosophy statement for general education with the addition of the attributes as graduation requirements for all students. My goal throughout this process was to make sure the three components of sustainability (global environmental awareness, social responsibility, and interpersonal and intrapersonal skills) were included among the attributes that finally passed the College Academic Senate. On June 19, 1998, this goal was achieved.

Positioning the Initiative So It Was the Least Threatening Choice

Once the attributes were passed as core graduation requirements, the College Academic Senate engaged in an additional two-year process of soliciting implementation models from the academic community and conducting two rounds of voting by all the faculty and academic administrators to select a model for implementation. This was a critical step because the implementation plan for changing the general education requirements had to be passed not just by the College Academic Senate, where there was a higher percentage of innovators and early adopters, but by the entire college faculty of more than 250.

In previous general education requirements, students were required to take certain distribution requirements for a degree (e.g., eight credit hours of humanities, six credit hours of social science, one science lab, one math). In the general education review, some people wanted to change these distribution requirements or do away with them completely. But the college faculty had a deep investment in the existing requirements. Faculty salaries are derived from the number of students they teach; we knew if we threatened the faculty's salaries by lowering enrollment in required courses, we would lose the entire general education revision.

Those in support of the attributes made a strategic decision to push for a model that did not change the existing distribution requirements so we could position it as the most acceptable model, creating a win-win with the territoriality issue. A small but politically well-positioned group of us described why this model produced the least threat to the course loads and the salaries of the faculty, helping produce the votes we needed. Instead of adding threatening new courses in this model to cover the attributes, disciplines integrated the desired attributes into the existing courses. Honoring the existing curricular structure was essential to gaining support from the faculty as a whole.

Support by the old guard faculty was crucial, since they were powerful in both the College Academic Senate and our strong faculty union. The head of the senate was powerful and articulate, capable of making or destroying initiatives at his will. I chose to meet with him early on in the general education review process. I acknowledged his power and mentioned that I knew he was creating a legacy that would remain after he retired. I told him what we were attempting to do and asked for his help in getting it implemented successfully, suggesting this could be part of his legacy. This approach could have backfired, giving him the ammunition he needed to cause a lot of trouble, but it did not. He may have responded positively to the legacy argument (to this day, I am not sure), but he did realize this was a potentially enormous change for the college that could affect his department (political science) and his student load. He became a member of the General Education Committee. When we asked the academic community to submit models for implementation, he authored the winning model, which both protected his discipline and implemented the attributes.

Key Implementation Strategies

A key strategy used by the supporters of the attributes was to include the senate attribute approvals in a report written for the North Central Accreditation reviewers in 1997. As a result, the accreditation agency will be looking in future reviews for evidence that we are implementing the attributes, thereby making it harder for someone to derail them.

As a member of the committee, I made sure the implementation steps for all the models contained two important pieces: inclusion of the attributes on student transcripts, and a statement that any general education distribution courses that were not approved for at least one attribute within two years would drop off the general education distribution list. This last statement created a tremendous incentive for faculty to get their courses approved for attributes, since they wanted to stay on the general education distribution list to protect their enrollments.

Once the model and implementation plan was passed, the ad hoc General Education Committee used existing faculty development structures, such as required staff development days, discipline meetings, and department meetings, to identify who within each discipline was responsible for implementation. We gave both the full-time faculty and the adjunct faculty assistance through workshops, one-on-one help, and Web development <http://www.oaklandcc.edu/assessment/gened/index.htm> of sample learning objectives, learning activities, and assessments for each of the attributes. We also assisted faculty by helping them with the application to their courses and supporting them at the College Curriculum Committee meetings. Upper administration provided course release time for this faculty-to-faculty assistance, at the committee's request. For the first year of implementation, a campus general education coordinator at each campus plus a college-wide general education coordinator received one class of release time in both the fall and winter terms. In the second year, the college-wide coordinator received one-quarter release time. To receive this release time, we created lists of what still needed to be accomplished for effective implementation, which we presented to the administration.

At OCC, faculty members must attend twenty hours of faculty development each year or lose some of their salary. We convinced the administration that attribute workshops should be counted for faculty development credit. This credit incentive, in combination with the need to get attribute approval to stay on the general education distribution list, increased faculty attendance at the attribute workshops. At the workshops, there were outstanding interdisciplinary discussions about the attributes—how to define them, teach them, and assess them. Over seventy full-time and many more adjunct faculty attended workshops about sustainability-related attributes. Through the staff development days, the entire full-time faculty received information on all the attributes, including environmental literacy and social responsibility. Within two years, over two hundred courses had been approved for attribute coverage.

While faculty chose to incorporate at least one of the sustainability attributes in over seventy-two courses, only twenty-five courses include the global environmental awareness attribute. This did not surprise me; I knew environmental offerings were a weakness in the curriculum. However, the structure of the general education requirements provides an avenue for correcting this situation. In fall 2002, we started another round of outreach to faculty. During our mandatory staff development day, some of us showed how few courses include the attribute "to understand the global environment." We explained to the faculty that there is an opportunity to gain more students by including the attribute, and we are available to provide assistance. We already have it as a degree requirement; now we want it to be a stronger component of the curriculum at the college.

Closing the Loop: Curriculum Review

At the same time that we finished the environmental scan, the innovators at the college sought to institutionalize the process of reviewing curriculum to check for concurrence between the college catalogue and classroom practices and to motivate academic disciplines to incorporate cutting-edge developments in their fields within the course offerings. Once the general education attributes were passed, we made sure to include a section in the review process on how the attributes were being taught and assessed. At one point, the College Academic Senate leadership asked us to back off on the formation of the review process because there was already so much going on. About five of us on the committee refused to listen to them (the joys of tenure), because we knew North Central Accreditation was coming, and it would be embarrassing to have no curriculum review process. We kept meeting and moving along, and the senate leadership finally came around and established the Curriculum Review Committee. This review committee provides a method for ensuring the ongoing implementation of the sustainability-related graduation requirements.

The Ultimate Outcomes?

The advantages of including sustainability education as part of the graduation requirements are numerous. New faculty and employers will see it as an institutional norm. It is a quick way (once passed) to infuse sustainability throughout multiple courses. It is a powerful statement to the students about the importance of some of the components of sustainability. The disadvantage is that general education gets reviewed periodically, and someone has to be willing to assess the emotional and intellectual climate at the college regarding this, go to the meetings, and keep the critical mass alive.

The attributes for global environmental awareness, social responsibility, and interpersonal and intrapersonal skills are now included in all the degrees in the college catalogue. Over the next ten years, as implementation continues and the next general education review begins again, even in the unlikely event that the worst happens and the attributes are changed, we have still accomplished a lot. Many faculty attended hands-on workshops about sustainability-related attributes, where they developed their own learning activities and assessments. Over seventy-two courses now contain at least one of the three sustainability attributes. Other colleges are using our changes as precedents at their institutions.

Parallel activities in energy efficiency, recycling, and futurist thinking at the college have helped build a momentum toward sustainability in all the sectors of the institution. The college goals for 2002–2007 adopted by the board of trustees in July 2002 state that the college should promote a global perspective and identify all courses that have the global (environment) attribute and consolidate information into an orientation handout booklet. I am even beginning to discuss with the chancellor signing the Talloires Declaration for the college. As with many other processes for societal change toward sustainability, vision and persistence are vital. What happens in the future is up to us.

References

Adler, Nancy, and Mariann Jelinek. 1986. "Is Organizational Culture Culture Bound?" *Human Resource Management* 25(1): 73–90.

Aguilar, Francis. 1967. *Scanning the Business Environment*. New York: Macmillan.

Cialdini, Robert Beno. 1984. *How and Why People Agree to Things*. New York: Morrow.

Kinnear, Thomas, and Kenneth L. Bernhardt. 1986. *Principles of Marketing*. Glenview, Ill.: Scott, Foresman.

Morrison, James. 2000. *Environmental Scanning in the Strategic Planning Process*. Chapel Hill, N.C.: College Board.

Rowe, Debra. 2000. "Motivating Students to Be Citizens Who Are Positive Change Agents for Career and Community Success. *Michigan Council for the Social Studies Journal* 12(1): 35–38.

III

Building Buildings, Building Learning Communities

8

Can Educational Institutions Learn? The Creation of the Adam Joseph Lewis Center at Oberlin College

David W. Orr

Oberlin College combines a four-year liberal arts institution, enrolling 2,300 students, with the Conservatory of Music, enrolling 600 more. The college is integral to the small town of Oberlin, Ohio (population 8,500), which is located about thirty-five miles southwest of Cleveland and is surrounded by farmland and suburbs.

Organizations that learn relative to the rapidly changing environments in which they exist, according to MIT professor Peter Senge, have three characteristics (Senge 1990; Senge et al. 1999). First, they are oriented to what people "truly care about," and are not focused on daily crises. In learning organizations, people build shared visions that require "unearthing shared 'pictures of the future' that foster genuine commitment and enrollment rather than compliance" (Senge 1990, 9). Second, conversations in learning organizations tend to produce "shared understanding, deeper meaning, and effective coordination." People do not just "talk at one another, engaged in never-ending win-lose struggles." The process of genuine learning, in other words, changes who has lunch with whom, as well as the content of what is said. Third, organizational learning requires the capacity to understand complex systems and to "see how their own actions and habitual ways of operating create their problems . . . and difficulties for people in other parts of the organization."

Organizations typically situate themselves relative to the competition for market share, political power, or influence. This can be inconvenient, however, if the entire herd is headed over a cliff. People in organizations capable of learning ask whether the game is worth playing at all. For the captains of the global economy, for example, it would be worth asking whether the world needs more clever ways "to sell more stuff to more people more of the time." Some of this stuff is lethal in parts per million. Some of it contributes to climate change and biotic impoverishment. Some of it causes obesity and human incapacitation of various kinds. Most of it is produced, packaged, and consumed wastefully. But I doubt that anyone in any legally chartered corporation really intends to kill their customers or the planet, even if by inches. Rather, I think they seldom ponder such things. Organizational learning at its best means rethinking what organizations do and how they do it relative to a larger standard of human and ecological health. Real organizational learning is not just a matter of reconfiguring the organization to do more efficiently and happily what should not be done in the first place. It is a deeper and more honest process of seeing patterns that connect what people in organizations do to and for people and their prospects elsewhere.

Specifically, can organizations that purport to advance learning themselves learn to recalibrate their mission and operations to the larger facts of global ecological change? The obstacles are significant. Higher education has tended to fashion itself into an industry beholden to other industries (Press and Washburn 2000) and is thereby complicit in larger societal and global problems. In Thomas Berry's words, we have fostered "a mode of consciousness that has established a radical discontinuity between the human and nonhuman" (Berry 1999, 4). And we take great pride in equipping our students to do well-paying work in an unsustainable economy—the rough equivalent of preparing them for duty on the *Titanic*. Many administrators and faculty acknowledge larger global environmental trends but have yet to adjust institutional behavior or curriculum accordingly.

What follows is a midcourse report of one learning experiment in higher education, with the caveat that it was not begun with any such intention. It was aimed, rather, at solving a practical problem in a small academic program. The possibilities for institutional learning came later as the project unfolded and revealed opportunities to develop additional operational and educational capabilities having to do with college buildings, landscape management, energy use, resource flows, and environmental impacts. But those opportunities raised larger questions about the purposes of the institution relative to the declining habitability of the earth as well as the transition, as Senge puts it, from colleges as "knowing organizations" to ones that function as "learning organizations" (Senge 2000).

Beginnings

In June 1995, the president of Oberlin College, Nancy Dye, approved an effort to build an environmental studies center to house a rapidly growing academic program. Although the project depended entirely on her interest and support, the initiative originated in the Environmental Studies Program (which I chair), not from the usual college planning process. To avoid taking money from what were regarded as higher priorities, approval depended on my raising funds from sources not otherwise likely to give to the college. I was given two years to raise what was originally estimated to be $2.5 million (but eventually grew to $7.2 million) and offered one course relief from my normal teaching load. The design effort was to be a collaboration between the college construction office (headed by an architect) and the Environmental Studies Program.

These initial conditions influenced the evolution of the project to the present. The fact that it was conceived and funded outside the usual channels was both a source of strength and a weakness. Being somewhat independent of the college bureaucracy at the outset, the project developed with more imagination than might otherwise have been the case. But that degree of independence came at a price: college buy-in has been inconsistent. The president's support did not necessarily translate into active support or even neutrality of other members of the administration, faculty, or trustees. The separation between the vision behind the project and institutional power—a schism between responsibility and authority—made the process awkward from the start. Because of its idiosyncratic nature, the project was vulnerable to the vicissitudes of college politics, making successful completion contingent on moving quickly. Constraints on the sources of money meant that the project would have to appeal to potential donors on grounds other than loyalty to the college. In other words, this would have to be aimed to attract support because it set a higher standard and was intrinsically interesting. The situation was paradoxical. Had we waited for the college to build an environmental studies center, we would still be waiting. But had the college undertaken to do it, the likely result would not have been very green. We began the endeavor in the hope that the institution would eventually take full ownership of it. And the project would have to be exciting enough to attract financial support but cheap enough to build—a middle ground between the Taj Mahal and a double-wide trailer.

In the summer of 1995, I made four decisions that shaped the design process. First, the programming phase would be open to students, faculty, and the wider community. In a world rapidly coming undone, we would use this project as an educational exercise in how we might stitch landscape, materials, energy, and water together in the context of a small building. I hoped, too, that participation would help to create an active constituency for the project. A second decision was to make the building an example of the highest possible standards of ecological architecture. No other building would be worth the effort anyway, but neither would any other kind of building be interesting to potential donors with no prior connection to the college. The third decision was to engage a team of ecological designers to work on this project, giving it a breadth of design integration as well as national prominence. Fourth, I asked John Lyle, a widely respected designer from California Polytechnic Institute, to facilitate the public design sessions—what architects call charettes. To help engage the campus community and coordinate details, I hired two graduates from the class of 1993, Brad Masi and Dierdre Holmes, as project assistants. Brad's good-natured, workaholic, and disheveled passion contrasted with Dierdre's cool, buttoned-down, incisive competence; both worked with imagination and energy.

During the fall and winter of 1995–1996, the building program emerged in a series of charettes in which some 250 students, faculty, staff, and members of the wider community participated. Three broad goals emerged. First, we decided to aim for a building that would cause no ugliness, human or ecological, somewhere else or at some later time. Like truth, beauty, and justice, that standard is beyond mortal attainment, but there is no other worthy standard. Second, we decided that the building and its landscape would be made active parts of the curriculum, not anonymous places where education just happened disconnected from place. We would aim to reconnect an increasingly disconnected urban clientele with soils, trees, animals, landscapes, energy systems, water, and solar technology. Third, we intended to use the project to develop and integrate a new set of analytical tools into the curriculum such as least-cost, end-use analysis, full-cost accounting, and systems analysis.

Many, I think, regarded this as a quixotic effort not likely to amount to much. We quickly confirmed their worst suspicions by engaging the finer points of the human condition such as alternatives to the modern propensity to mix drinking water with human excrement—a subject that much amused the sophisticated. Blinded by zeal, we proceeded nonetheless. In hindsight, the final program was both ambitious and foolhardy. We decided that the building would:

1. Be integrated with the curriculum.
2. Evolve with advancing technology.
3. Discharge no waste (i.e., drinking water in, drinking water out).
4. Use sunlight as fully as possible.
5. Use only wood from forests certified as managed sustainably.
6. Minimize use of toxic materials.
7. Be integrated with the landscape as a single design system.

In addition, the performance of the building energy and water systems would be made transparent to the public and evaluated by an authoritative agency independent of the college. Further, the landscape around the building would be designed around three questions: (1) Where are you located, and what does this place want to be? (2) Where are you in time? (3) What can nature and humans do in this place? Accordingly, the east side of the site would be developed as a wetland, pond, and small forest using native plants reflecting the biotic past of northeast Ohio. To the south, the landscape would be designed as a "sun plaza," featuring a large sundial marking the solstices and equinoxes. The north side would be a working landscape with an orchard and gardens built and maintained by Oberlin students under the supervision of David Benzing, a well-respected biologist on the college faculty.

In the fall of 1995, twenty-six architectural firms responded to our request for qualifications. We subsequently interviewed five and eventually selected the firm of William McDonough + Partners as the lead architect (McDonough and Braungart 2002). In contrast to most other college-architect relationships, this assignment required coordination of a larger design team, work with Oberlin students, and research on environmentally benign materials and construction methods. Design began in earnest in February 1996 and concluded with groundbreaking in the late summer of 1998. In contrast to several other projects around the country initiated at the same time, the Lewis Center was completed substantially as described in the building program developed in 1996.

I recall much of that time as a blur of long meetings, airports, and occasional crises, but four features stand out in my memory. First, even with the active support of the president, the project developed at the margin of institutional commitment and was therefore at risk from the beginning. Twice it came within a gnat's eyebrow of termination, first because of fears originating in the Development Office that I could not raise the necessary funds. In September 1996, the project was put on hold, but the following week two large gifts restored administrative confidence, and things were back on track. In the summer of 1997, the architects were ready to quit over what they regarded as aggravating behavior by the college facilities construction office. Following a lengthy and difficult phone conversation, they agreed to stay on, and the president subsequently resolved the problem. Once again we were back on track.

Second, the fear of failure colored the reception of the project among some of the senior administrative staff. In contrast to the president, some feared that this would be an embarrassing failure or that it might ruin the reputation of the college by identifying it with the lunatic fringe. A few, I suppose, feared that it might be too successful. And when the bullets were flying in the summer and fall of 1996, I recall seeing fewer folks around than were on the parade ground at the start. In this regard, higher education differs markedly from the nonprofit world where I had spent eleven years before arriving at Oberlin. There, the point is to decide what needs to be done, choose the smartest way to do it, and get on with it. But in the academy, where the clock speed can be glacial, entire careers can be organized, as philosopher Mary Midgley puts it, to "make no mistake." That may explain why the first response to proposed change often is a recitation of the nineteen reasons why it cannot be done with scarcely a nod to why it should be done.[1]

Third, the project was made more difficult than it had to be. In hindsight, I believe this is because it was highly subversive of operations and values that are typical among institutions of higher education. Colleges and universities are risk-averse organizations, yet we had embarked on a project that involved (or was perceived to involve) some risk. Colleges, like most other industrial age bureaucracies, are organized as separate fiefdoms. To create a building like the Lewis Center, however, required a high level of integration across divisions of curriculum, finance, operations, communications, admissions, and development—a level of integration that we did not achieve. We set out to design and build with an eye for the long term, but college and universities orient to shorter time horizons, particularly in matters of budget and finance. Colleges are hierarchically organized, but the energy for this project, as distinct from the support of the president, did not come from the top. Finally, from John Henry Newman (author of *The Idea of A University,* 1852) to the present, liberal arts colleges have deemphasized the practical arts, yet this was an effort to join theory and intellect with practical application.

Fourth, the design and hardware of a high-performance building, as complicated as those can be, were much easier than the human aspects of the process. When we stumbled, or nearly so, the cause almost always had something to do with human dynamics and most often the failure or refusal to communicate across the divisions of outlook, assumptions, rank, and officialdom. This was true between the college and the architects, as well as within the design group over practical and philosophical differences.

The Building

By January 2000, though not completed, the building was ready for "substantial occupancy," but what exactly had we done? We had certainly raised the bar for academic architecture, but we had also created a complex machine unlike any other building on this or any other campus. Conventional buildings are rather like manual typewriters that need periodic ribbon changes and oil. High-performance buildings are more like notebook computers that need software upgrades and networking capacity. Both typewriters and computers produce paper covered with symbols, but there the similarity ends. The Lewis Center has better technology, more complex controls, and higher performance potential than conventional buildings. But performance remains potential until the complexity of building systems is mastered. For this reason, high-performance buildings go through a complicated commissioning process to check out systems and evaluate how well performance matches expectations, followed by a period of modification and adjustment. As sophisticated, complex systems, they require a substantial increase in the capabilities of the people charged with their maintenance and operations. The Lewis Center was commissioned in the summer of 2000 by the facilities management firm contracted by the college. A Texas firm specializing in energy efficiency did a separate analysis. Given different priorities and assumptions, the reports differed substantially.

We had modeled the building as it evolved through design, but to save money, the college facilities managers decided not to run an energy simulation, known as DOE-II, on the final construction documents. Had we done so, we would have discovered that an electric boiler, intended as an emergency backup system, had become the primary heat source for the atrium at a substantial increase in energy use. Other corrections included undoing some of the "value engineering" changes that were carried out late in the design process, purportedly to reduce costs. Value engineering, I learned, is an expensive and arcane way to reduce value without necessarily improving the engineering or anything else.

Counting on the rapid development of distributed energy systems such as photovoltaic technology and fuel cells, the Lewis Center was designed as an all-electric building. Accordingly, the building included a 4,700 square foot solar photo-voltaic (PV) array rated at 59 kilowatts with the possible later addition of a "regenerative" fuel cell that would store electricity in the form of hydrogen and provide electricity by recombining hydrogen and oxygen at night. The photovoltaic array was installed and operational by the fall of 2000. Since electricity is expensive to store and because Ohio is a net-metering state, we connected the PV system to the electric grid, selling excess power to the grid on sunny days and buying power back at night and cloudy days at the same rate. Building energy performance for the first year of data authenticated by the National Renewable Energy Lab (March 1, 2001, to March 1, 2002) shows that the Lewis Center used 27,000 Btus per square foot compared to a national average for new classroom or office buildings of around 80,000. Normalized for climate, the best comparable building in the same period used approximately 35,000 Btus. Eleven percent of the energy in the Lewis Center is required to run the waste processing system; another 3 percent powers exterior lighting. Subtracting the PV production from the gross use, the net site energy use is 11,600 Btus or about one-third of what is the next best comparable building on a college campus. Changes made in the spring of 2002 will lower building energy use further by an estimated 25 to 30 percent.

The Living Machine to process wastewater became operational in the summer of 2000. The system was designed to handle 2,300 gallons of waste per day, but throughput from toilets, urinals, and sinks turned out to be much less. By the spring of 2002, the system was producing water that exceeded federal tertiary standards using only plants and animals, much like a natural wetland. The landscape was developed in stages. The forest, wetland, and pond, featuring plants native to northeast Ohio, and an orchard were planted in the summer of 2000. Gardens maintained by students were planted in the summers of 2001 and 2002 under the supervision of biologist David Benzing.

Aside from the technicalities, the Lewis Center caused different reactions around the college and beyond. College politics, it is often noted, are sometimes so nasty because the stakes are so incredibly small. All of us in the project, the president included, were variously applauded and criticized. It goes with the territory, as they say. On the dedication day, for example, I received an e-mail from an angry faculty contrarian threatening to expose the project as a fraud. Another local climate change skeptic attempted to convince the donors that the Lewis Center was a left-wing boondoggle. Some in the administration were reportedly much amused by the Living Machine to process our wastewater, a landscape that did not resemble that of a country club, our unfamiliar solar technology, and the challenges of ecological design. But ideas, it is said, proceed from opposition to ridicule to acceptance as merely obvious.

Misinterpretations of the project were also common. An e-mail sent by an earnest young woman, for example, excitedly asked for a tour of the "Oberlin poop building"—the one, as she put it with great admiration, "powered by human feces." For its part, the press only sometimes got the story right. The *Chronicle of Higher Education* (June 21, 2002) described the building as reflecting larger financial problems of the college in the bear market of 2002 and as failing to meet expectations about energy use and wastewater systems. Wrong on all three counts, the reporter said he added this to lend some controversy to an otherwise dreary article.

The upshot? Beware of those with axes to grind and time to burn. Be aware of what can go on behind closed doors. Be instructed by the astonishing power of the rumor mill to both inflate and deflate or create something entirely novel. And be humbled by the manifold ability of the press to get complicated things wrong. But most important, be true to the vision.

Evaluation

The fact that the Lewis Center was completed substantially as described in the building program and performs largely as hoped is a tribute to the commitment and stamina of the donors, president, the college staff and faculty who worked on the project, and the architects and design group. The building has won three architectural awards, two awards for construction techniques, and one for energy efficiency, and it has been recognized by the U.S. Department of Energy as one of thirty Milestone Buildings in the Twentieth Century (Malin and Boehland 2002).[2] It continues to attract a high level of attention in the national media as a pacesetter in architecture (Petersen 2002). The building's tens of thousands of visitors include representatives from several hundred colleges and universities, federal agencies, and private companies. More important, the Lewis Center has helped to stretch the ecological imagination and competence of Oberlin college students in fields of solar technologies, ecological engineering, horticulture, landscape management, and ecological design. In retrospect, a number of lessons can be drawn from our experience that are instructive for others intending to build high performance buildings.

Make the Building Program Fit the Fundraising Strategy

Raising money for buildings can be a hard job, especially if one sells only a building. If this experience is a useful guide, the ideas embedded in the building program are more important than the stated need for the building itself. The more obsolete the ideas, the harder it is to create enthusiasm about the project among potential donors or anyone else for that matter. Buildings are means, not ends, but a means to what? The Lewis Center was conceived as an experiment in the application of solar technology, ecological engineering, products of service, ecological landscaping, sustainable forestry, and the art and science of ecological design. The larger goal was to better equip our students to solve twenty-first century problems. This will require significant changes in how we think about buildings and their larger upstream and downstream effects over the long haul. This is both daunting and exciting, but if we intend to stay around awhile longer, it is absolutely necessary to rethink the built environment as a keystone of a sustainable world. Good ideas, in other words, tend to attract money.

Thoroughly Integrate the Design Process

In the making of the Lewis Center, the design team was not as well integrated as it should have been. As a result, an ambitious building program and visionary design were not sufficiently calibrated with the engineering. Part of the difficulty lay in the fact that in the mid-1990s, the talent necessary to design high-performance buildings was not available locally. As a result, we assembled a team that included a dozen or more people scattered throughout the United States, which raised costs and made coordination difficult. More important, the mechanical engineers did not entirely share our vision of high-performance, low-energy design. The resulting lack of integration of engineering with the overall design goals proved to be the weakest part of the building design.

Maintain Creative Flow

After participating in one of the early design charettes, one faculty veteran described the event as the most exciting he could recall in his time at Oberlin. The excitement of the first months, however, diminished by the end of 1997, largely because of the slow pace of college decision making. What should have taken a year or so to design was extended for thirty months, impairing the creative flow of the design process and the morale of those involved. Midway through design the administration hired an outside firm to manage the college physical plant, including oversight on this project, further impairing the continuity of vision guiding the project.

Develop a Larger Learning Process

This project originated and has to this point remained on the periphery of institutional consciousness. No formal or informal feedback loops bridged this project with other building projects or to institutional operations. Other than the president, the project had no strong advocate within the administration, which may explain why no effort was made to develop a shared vision—what Senge calls "common mental models"—

among trustees, senior staff, facilities management, and faculty. After commissioning, the administration initiated no review of the project with all the participants to determine what worked well and what did not.[3] As a result, different and somewhat antagonistic views of the project and of the design process exist between the college administration and the faculty and design group that worked on the building. It is fair to say that the Lewis Center does not yet reflect a deeper institutional commitment to sustainability, energy efficiency, the transition to solar power, ecological restoration, and biological diversity that were central to both the building program and the Environmental Studies Program. On the contrary, the project has been tacitly regarded as an isolated experiment, not as the beginning of a larger change. A new science facility, begun a year after the Lewis Center, has few, if any, environmentally redeeming features and commits the college to substantially higher energy costs than necessary. Two years after commissioning, a member of the design team observed that "our story truly isn't their story." Perhaps in time this will change.

Account for the Life Cycle Costs and Collateral Benefits of Buildings

The cost of a building is often confused with the initial price of the thing, leaving out life cycle and environmental costs. As a result, institutions often get cheap buildings that come in on budget but are expensive to operate and environmentally destructive. The total project cost of the Lewis Center, $7.1 million, includes a building endowment, design fees, research, and construction. In simple math, the construction cost of $4.8 million divided by 13,700 square feet gives a cost of $350 per square foot. But a more accurate "apples to apples" rendering requires subtracting $1.2 million in unconventional costs, including those of relocating sanitary and storm sewers, and the costs to construct an oversized parking area. In addition, since most buildings do not include a sewage treatment facility or a power plant, valid comparisons require subtracting the costs of the Living Machine and photovoltaic array. A valid comparison, in other words, shows that the square foot cost of the basic building falls between $250 and $260, not out of line with construction costs of classroom or office facilities of the same size built in northeast Ohio at the same time. But focusing on construction costs alone misleads.

A full assessment of costs would include those to operate and maintain the building over its lifetime, as well as its environmental impacts. Further, costs stand in relation to benefits. In this case, the collateral—and mostly unaccounted—benefits to Oberlin College include a substantial amount of national publicity, increased student yield, increased donor interest in the college, and a facility that enlivens the curriculum in environmental studies and attracts a rising level of student interest.

It Ain't Over 'Til It's Over

On completion, most buildings have maintenance and depreciation schedules. The Lewis Center, in addition, was intended to have a learning trajectory as a building that would evolve over time as technology and management skills improved. Can buildings in northern Ohio reliably generate more power than they use? Can they foster biological diversity? Can building operations become a form of lasting educational value? Building owners, on the other hand, typically expect the product to look good, work reliably, and be finished once and for all at the time of occupancy, not as something to be tweaked, tinkered with, studied, and improved. Regarding buildings as evolving, not fixed, assets requires a longer view, patience, growing technological skill, and an ecological vision of the built environment. It requires institutional stamina in the pursuit of a larger vision of a sustainable and decent human future.

Follow the Logic

The Lewis Center is a very small but potentially important piece in a larger process of learning how we might reduce or eliminate the environmental impacts of buildings. The construction, maintenance, and operation of buildings of all kinds represent 40 percent of our total raw materials and energy use, 40 percent of our sulfer dioxide and nitrogen oxide pollution, 33 percent of our carbon dioxide emissions, 25 percent of our wood use, and 16 percent of our water use. Thousands of college and university buildings, existing and planned, are both part of the problem and potentially a great opportunity to do something better. Imagine pollution-free campuses powered by sunlight. Might it be possible for colleges to aim to become climatically neutral within the next few decades? While Congress and the White House dawdle, educational institutions could begin to chart a different future as models of ecological design that equip students with the means to solve twenty-first-century problems. To examine this possibility, we hired the Rocky Mountain Institute to study carbon emissions on the Oberlin campus and develop scenarios showing alternative pathways to climatic neutrality by 2020. Intended to initiate a larger dialogue, the report at this writing remains officially under study.

Build a Bigger Story

The making of the Lewis Center helped to spawn a number of other initiatives and organizations beyond Oberlin College. In 1998, for example, with some initial financial support from the Environmental Studies Program, a highly entrepreneurial recent alumnus, Sadhu Johnston, established the Green Building Coalition in Cleveland as a way to move ecological design ideas from this project into a rust belt city. Directly and indirectly, the coalition has influenced dozens of other projects throughout the region, including the green renovation of an abandoned downtown bank building. Brad Masi subsequently went on to start a 70-acre Oberlin community–supported farm to market produce locally and raise awareness of food and land use issues. From a private reading course on ecological design, another group of students launched an organization aimed to help revitalize the city of Oberlin. Three other students purchased an expired car dealership and intend to build a multipurpose green building to anchor a critical part of the Oberlin downtown.

Plan For and Celebrate Success

The difference between success and failure is often only the stubborn refusal to fail in the face of daunting odds—more a matter of will than of intellect. Success begins by envisioning success and planning for it. In a team setting, momentum toward a successful conclusion is built and maintained by competent professionalism and a psychology of encouragement, appreciation, and generosity. And the difference between a good outcome and a great one is built into the personal dynamics that let a vision grow to its full stature or stop it short.

But there are many barriers to success. The ethos of an organization can inhibit the capacity for success. Institutions have collective personalities of sorts, and Oberlin College is no exception. Rooted in a Congregationalist church past, ours is rather understated. A clue, perhaps, is found in the collection of portraits hanging on the walls of the conference room in the administration building. The faces of somber, bearded, and fiercely righteous men glare down on their far more secular descendants. But if now more secular, we are hardly less serious than they about our various causes. But whether theological or secular, fundamentalism of any kind (or any-ism, including environmentalism) can inhibit institutional learning.

Other barriers to success originate in hierarchical authority that is, "a poor vehicle to cause imagination, commitment, passion, patience, and perseverance" (Senge 2000, 294–295). In contrast, learning organizations recognize and overcome "the numerous ways and means that [the institution] uses to squash innovation and force conformity" (Birkeland 2002,). And the ways to chill creativity and impede learning are many. Learning organizations manifest a different psychology that celebrates initiative and risk taking, and when mistakes occur, they practice "real forgiveness [and] reconciliation" (Senge 2000, 300). Learning within any organization, in other words, occurs most easily in an atmosphere of easy camaraderie, optimism, support, gratitude, and openness.

Conclusion

As an early version of a high-performance building, the Lewis Center, to paraphrase Wes Jackson of the land Institute, is equivalent to the experiment in flight at Kitty Hawk; we are ten feet off the ground. Someday others may design and build far better buildings that regenerate natural capital—the equivalent of 747s. To see our efforts as a small step in a longer view helps to restore perspective that is easily lost in the sheer excitement and busyness of designing and building. The story of this project will soon be forgotten. What should not be forgotten is the art and science of making the human presence in the world in a way that honors and protects the prospects of our descendants for as long as we can imagine.

Buildings, in this perspective, are means, not ends. The Adam Joseph Lewis Center was initially conceived as a solution to a facilities problem and as a laboratory for ecological design to help equip this generation of students to carry on what our generation has only begun to do (Orr 1994, Orr 2002). They will have to stabilize and quickly reduce greenhouse gas emissions from 8.5 billion tons to around 3 billion tons, stop the loss of biodiversity, reduce population growth, rebuild cities, eliminate waste, learn how to grow their food and fiber sustainably, and radically improve fairness within and between generations. They will have to reshape economies and public institutions to fit ecological realities—

what Thomas Berry (1999) calls the "Great Work" of the twenty-first century. It is nothing less than the recalibration of human intentions with the way the world works as a physical system. Ecological design will help buy us some time, but that is all it can do. The question is, Time for what? If ecological design is used to rationalize and support the most reactionary parts of the global consumer economy, we will have lost the best chance we will ever have to build a genuinely sustainable, fair, and spiritually sustaining civilization. If used to build communities and entire nations rooted in equity, decency, ecological competence, and compassion, that time will have been well spent.

Can institutions of higher education become learning organizations? On Mondays, Wednesdays, and Fridays, I am inclined to think so. On alternate days, I am not so inclined. The barriers to organizational learning are found in inertia and the license sometimes given to the lesser side of human nature. Learning organizations, by contrast, are institutions energized by the angels of our better nature: a sense of good possibilities, optimism, human openness, shared vision, and awareness of our connectedness in space and time. Educational institutions committed to the real work of building a sustainable and decent human future and willing to learn what that requires of us would be exciting and challenging places. More to the point, they would equip the rising generation to see that the world is rich in possibilities and prepare them to act competently in that light.

Notes

1. The Danish Work Psychology Department at the Technological Institute of Denmark has assembled a list of "Proven and Effective New Idea Killers," including: Let's think more about that, LATER. I know it's not possible. We are too small/big for that. We have already tried that, that will be too expensive! That will mean more work. We have always done it this way, so why should we change now? Let somebody else try it first. We have no time for that. It sounds fine in theory, but how will it work in practice? We are not ready for this idea yet.

2. The building has received these awards: American Architecture Award from the Chicago Athenaeum, Honor Award from the AIA Committee on Architecture for Education, Build Ohio Award from the Associated General Contractors of Ohio, Build America from the National Convention of the Associated General Contractors of America, AIA Committee on the Environment, 2002, and A Governor's Award for Excellence in Energy Efficiency, 2002.

3. I organized a retrospective on the project with the design team on August 13–14, 2002.

References

Birkeland, Janis, 2002. *Design for Sustainability*. London: Earthscan.

Berry, Thomas, 1999. *The Great Work*. New York: Bell Tower.

McDonough, W., and M. Braungart. 2002. *Cradle to Cradle*. Washington, D.C.: North Point.

Malin, Nadav, and Jessica Boehland. 2002. "Oberlin College's Lewis Center." *Environmental Building News* 11 (July–August):7–8.

Orr, David W. 1994. *Earth in Mind*. Washington, D.C.: Island Press.

———. 2002. *The Nature of Design*. New York: Oxford University Press.

Petersen, John. 2002. "Appraising Success of the Adam Joseph Lewis Center." *Environmental Program Newsletter* (Spring).

Press, Eyal, and Jennifer Washburn. 2000. "The Kept University." *Atlantic Monthly* 285 (March): 39–54.

Pulley, John L. 2002. "Well-Off and Wary." *The Chronicle of Higher Education*. June 21.

Senge, Peter. 1990. *The Fifth Discipline: The Art and Science of the Learning Organization*. New York: Doubleday.

———. 2000. "The Academy as a Learning Community." In *Leading Academic Change*. Edited by Ann Lucas and Associates San Francisco: Jossey-Bass.

Senge, Peter, et al. 1999. *The Dance of Change*. New York: Doubleday/Currency.

9

The Development of Stanford University's Guidelines for Sustainable Buildings: A Student Perspective

Audrey B. Chang

Stanford University, a comprehensive private research university, is located in a suburban community near Palo Alto, California, about thirty miles south of San Francisco. It enrolls 6,600 undergraduates and 7,500 graduate students. The campus spans over 8,000 acres, two-thirds of which is open land or only sparsely built.

In March 2002, Stanford University published *The Guidelines for Sustainable Buildings*, the result of a joint effort of students, faculty, and senior staff to integrate sustainability into the design process for its central campus buildings. This landmark achievement is particularly notable in that student interests lie at the roots of this effort. Developing the guidelines was a chance for students to apply their classroom knowledge to play an active role in helping to change the world, starting with Stanford. The past few years of working with the Stanford administration brought tremendous insights, seemingly impossible roadblocks, and mind-bending frustrations.

Through the process leading up to the release of the *Guidelines*, I sat on both sides of the table—with activist students and with cautious administrators—and learned from both. I am a 2002 graduate of Stanford, where I completed an M.S.E. in energy engineering, an individually designed program in the School of Engineering, and a B.S. in earth systems, an interdisciplinary environmental science major. In my last two years at Stanford, I made the *Guidelines* and green building at Stanford a personal quest. I was intimately involved in the process leading to the release of the *Guidelines*: first as a member of the student group that initially convinced the Stanford administration to examine the sustainability of its buildings, then as a student representative on a committee charged with developing the *Guidelines*, and finally as a lead author of the *Guidelines*. I never expected to be so closely involved.

Students for a Sustainable Stanford

Through classes and my personal interest, I learned of a growing number of schools, cities, and businesses that construct their buildings in an environmentally sensitive fashion. Repeatedly, I turned toward my own campus and wondered why Stanford was not also building sustainably. I was not alone. But all of the student movements in the past, mostly among undergraduates, were short-lived, hampered by the quick turnover of students at the university. Most students did not take action until their upperclassmen years, and any progress made soon stagnated as the students graduated and pressure on the administration diminished.

In spring 2000, I started running into a few other students who also individually had begun to ask why the university's buildings were not built to be green. But momentum did not pick up until the start of the following fall, when five of us, none previously friends, gathered to discuss our plans for definitive action. Two of us were law students intent on reducing Stanford's greenhouse gas emissions, and the other three were undergraduates (two undeclared sophomores, and one earth systems senior—me) interested in green building at Stanford. Especially concerned by the numerous new buildings slated for construction in the next decade in Stanford's General Use Permit, we decided to join forces to push for a green building policy on campus.

We called ourselves the Stanford Task Force on Sustainable Building, and later renamed ourselves the Students for a Sustainable Stanford (SSS). We slowly recruited other students through word of mouth and e-mails to campus lists. Our meetings were open to anyone who was interested, from a one-time stop-in to active involvement. Membership in our group fluctuated with varying demands on people's time, but a core group of about eight students remained. Holding weekly meetings in borrowed space in the windowless basement of the law building, we discussed our strategy. From the beginning, we had little administrative structure to our group (it was not until a year later that we had our first coordinator). Individuals volunteered for certain tasks we felt needed to be done. Not everyone showed up every week, but e-mailed meeting summaries kept everyone up to speed. At key moments (such as meetings with administrators), everyone would coalesce into a cohesive group of students.

Gathering Our Facts

Our first step was to educate ourselves, and we spent as much time as we could spare doing research. Although some of us were knowledgeable about green and energy-efficient building practices, others within the group were not. We scoured the Web and talked to as many people in the field as we could. We recognized that these resources would be crucial to maintaining the group's continuity as older members graduated and new members joined and had the foresight to maintain a library of our findings and meeting summaries.

We also wanted to understand the administration and the process by which buildings on campus were designed. To do so, we unearthed the mysterious building production process at Stanford. There are 678 major buildings on Stanford's campus, encompassing 12.6 million square feet.[1] Land and Buildings is the branch of the Stanford administration charged with land use planning, landscaping, campus circulation, project management, and infrastructure planning and maintenance. The associate vice provosts of each of the four Land and Buildings departments (Capital Planning and Management, University Architect/Planning Office, Facilities Operations, and Environmental Health and Safety) report to the vice provost of Land and Buildings, who then reports to the university provost.

Along the way, we also learned (to our pleasant surprise) about Stanford's current environmental building practices, which already excelled among colleges and universities. As required by California law, Stanford in 2000 had diverted 50 percent of its waste from landfills. Stanford's Energy Retrofit Program (ERP) is another example. In the late 1990s, Stanford participated in the Environmental Protection Agency's Green-Lights program and converted all campus lighting fixtures to more energy-efficient T8 lamps with electronic ballasts. Although the effort cost $1 million and extended over four years, Stanford finished ahead of schedule and the benefits paid off: energy savings were visible on the campus level.[2] Stanford's only failure in these areas was not publicizing its efforts more.

Making the Case for Sustainable Buildings

Although these Stanford achievements are certainly commendable, we wanted more: a broader sustainability-based approach to redefine how Stanford constructs all of its buildings. Buildings for the large part are not designed up front to be as energy efficient as they could be. A few past attempts at green or energy-efficient buildings were viewed as failures due to unassociated reasons, and administrators tended to say, "Tried that, done that." One promising development was a new green field station being designed for Stanford's Jasper Ridge Biological Preserve, which has since been completed. Still, Stanford had never tried to implement sustainable strategies across all its buildings.

Initially, we attempted to recruit a faculty member to help champion our cause. After approaching several different professors, however, we found most were already overloaded and that none was willing to take on a completely new project. Even so, many expressed their willingness to serve as consultants, and we periodically returned to several faculty, as well as outside experts and alumni in the green building field, for advice.

To raise general awareness and support for these issues on campus, we mounted a large campaign to educate students, faculty, and administrators. We published op-eds in the campus newspaper and held open information sessions. Over twenty professors signed an initial letter of support. We successfully lobbied both the Associated Students of Stanford University (ASSU) and the Graduate Student Council to pass statements of support. Both of these organizations delegated their negotiating authority to our task force on these issues so that we spoke for over 14,000 students when advocating our proposals. We also set up a Website <http://sustainability.stanford.edu> and an e-mail list to which anyone could subscribe, now with over 100 subscribers.

Even with these various cross-campus endorsements, we knew we had to be prepared to convince the administration to take us seriously. Although we were united in a common interest in improving Stanford's environmental performance, we knew that an environmental argument by itself would not be enough to convince the administration to build green. We had to learn how to speak business. We realized that we had to present this as a win-win situation, making the argument that there would be so many secondary benefits that the administration could not reject our proposals.

We held that while remaining sensitive to economic concerns, it is possible to construct buildings that are healthy for their occupants and have minimal impact on the environment. Our central argument was that green and energy-efficient buildings would save Stanford a significant amount of money by reducing operating costs over the life of the buildings. As with any other institution or business, monetary concerns are a primary consideration of the administration. The productivity of researchers, students, and staff could be expected to increase in green buildings as well, supporting the academic mission of the university. Stanford could leap to the front of the pack in green building among other universities and in the industry.

Not only did we have to speak business, we had to act it too. We knew the chances we had to speak to senior administrators were few and far between for students. Before each meeting, we carefully planned our agenda and method of approach. We prepared professional PowerPoint slides and rehearsed our presentations. For each meeting, we temporarily exchanged our shorts and teeshirts for slacks and collared shirts. We knew we were coming from a disadvantaged position, and good impressions were necessary; a professional appearance was part of it.

When we first brought our concerns to the senior administrators, it was no surprise that they did not embrace the issue immediately. But we forged ahead. To make our case, we gathered more supporting evidence and slowly gained credibility. During the 2001 academic year, we met several times with the top building administrators (the vice provosts and associate vice provosts in Land and Buildings) and began a process with them to identify challenges to incorporating more green building procedures and techniques in campus construction. We also met with the university president, provost, and assistants to the president. In addition, we communicated with several members of the board of trustees, targeting two who strongly supported environmental causes. We contacted the undergraduate student representative on the board of trustees' Committee on Land and Buildings, and in spring 2001, he gave a presentation on the merits of green building to that committee.

I believe that the SSS was successful in gaining the ear of the administration for four primary reasons. First, our student group was diverse with representation from various fields (anthropology, biology, business, drama, earth systems, economics, engineering, international relations, law, and urban studies). Second, we were persistent and showed staying power. Membership in the group ranged from freshmen to graduate students, and the administration dealt with familiar faces throughout our campaign. Though some members have graduated or moved on, others remain to continue the efforts. Third, we concentrated on personal contact, believing it would be most effective, though more time-consuming, than correspondence simply through e-mail and telephone calls. Whenever we had a large statement to make or an op-ed piece published in the campus paper, members of the group traveled around campus to hand-deliver copies to all of the important administrative players. When we wanted a meeting with an administrator, we sent a representative to the appropriate office to ask for a meeting in person. Finally, SSS took a professional and cooperative approach. We did our research and went straight to the appropriate decision makers and emphasized that we wanted to help Stanford. The learning process went both ways: we learned about the Stanford building process, and the building administration learned about the benefits of green building.

Playing Politics

Throughout SSS's history, we tried to be sensitive to the administration's concerns. To drum up support for our movement, we were wary of collaborating with advocacy groups from the surrounding community. At the start of our campaign, Stanford was in the midst of a bitter and much publicized dispute with the local community over Stanford's General Use Permit and the use and development of the Stanford foothills. We knew Stanford would not be willing to enter another high-profile battle that pitted the environment against growth and Stanford against the surrounding communities. We could tell that administrators were wary of a protracted political effort and thus of any environmental advocates.

To avoid the "Stanford versus the world" mind-set and to circumvent any possible misunderstandings, SSS ultimately decided to keep the campaign within the Stanford community (students, staff, faculty, and administrators) and to work within the system. This stance has been a heavily debated one throughout SSS's history, as student frustrations frequently raged after seemingly nonchalant responses from the administration. Several times, we debated the benefits and disadvantages of resorting to public protest. Ultimately in each case, however, we decided that we did not want to elicit a confrontational response from the administration that could break down communications altogether. Instead, we took a more aggressive stance into the meetings with administrators, stepping up our pressure with words and more memos and more presentations. That tactic—playing the game like the administration was accustomed to—seemed to work in our case. It did not seem that other students had ever tried to do the same, and I think we gained respect in that fashion.

Formation of the Environmental Stewardship Management Group

In March 2001, six months after SSS was formed, the top Land and Building administrators proposed the creation of a new committee within their administration to review past projects and make green building policy recommendations. Unsatisfied with their proposal and doubtful that the committee would do anything besides add to the bureaucracy, we issued a counterproposal. Negotiations resulted in the formation of the Environmental Stewardship Management Group (ESMG) and a verbal promise to develop Stanford-specific sustainable building guidelines by the end of the calendar year. Chaired by the vice provost of Land and Buildings, the committee would comprise Land and Buildings staff as well as faculty and student representatives.

Given that many of the ESMG meetings were to be held over the summer and into the next school year, we faced the problems of ensuring continuity. Most administrators but few students or faculty remain on campus during the summer break, and it was difficult to find faculty who were willing to commit to regular meetings with the ESMG. Ultimately, we found one civil engineering professor and a member of the academic staff to represent the faculty. We could commit only two students to the committee who would be available for at least part of the time. One student, an earth systems junior, could attend only the summer meetings since he went abroad during the fall. I was the other student but was gone periodically over the summer. Although I was already in my senior year, my earlier decision to stay another year at Stanford to pursue a coterminal master's degree guaranteed that I would be able to continue working with the committee in the fall.

Although faculty and student representation in the ESMG was less than ideal, with administrators outnumbering faculty and students by more than two to one (we had originally proposed a three-way split of administrators, faculty, and students), it was the best we could come up with. We were comforted by the participation of a few administrative staff sympathetic to our cause and were buoyed by the progress shown by the creation of the committee.

Its membership determined, the ESMG met for the first time in May 2001. The students proposed a mission statement for the group that would ensure that a product would result. After a bit of tweaking the statement, the group agreed on its mission: "To develop a set of Stanford-specific sustainable building guidelines and integrate them into the facilities planning, design, and operations processes." The ESMG met biweekly through December 2001. A breakout working group, of which I was a member, met during the off weeks to hash out the details of the developing guidelines and reported back to the ESMG.

To LEED or Not to LEED?

The first contentious issue that the ESMG tackled, first raised in discussions between SSS and administrators, was whether to adopt Leadership in Energy and Environmental Design (LEED), the U.S. Green Building Council's green building guidelines and rating system. LEED is quickly becoming the universal standard for green building, within the United States and internationally as well. A number of other colleges and universities, including MIT, Emory, and the Los Angeles Community College District, have used or adopted LEED for their new construction. So why not Stanford? It also seemed to make sense for Stanford to take advantage of the research and investment that went into developing LEED. Administrators, however, were resistant to any sort of standard to which Stanford would be held.

Although I was disappointed that Stanford was unwilling to adopt a preexisting guideline, I was encouraged that it was willing to develop its own. Eventually, the *Guidelines* were structured into three main sections: process, technical guidelines, and decision-making tools. In an evaluative task, the working group reviewed the entire LEED guidelines and compared them with current Stanford building practices. We discovered scattered examples throughout campus of sustainable building practices (though not to the extent required by LEED). More could be done to apply them universally across campus. We also found some ways in which LEED is ill suited for Stanford. Some LEED points, geared toward traditional urban development, are not directly relevant to the building conditions at Stanford.

Next, the working group undertook a comprehensive review of various other green building guidelines and sustainability initiatives in existence among other schools as well as local governments. These included the City of New York High Performance Building Guidelines and the Minnesota Sustainable Design Guide. We drew inspiration from these sources, as well as LEED, and followed the format of all of these guidelines. We divided the technical guidelines by sustainability category: site design and planning, energy use, water management, materials, resources and waste, and indoor environmental quality. Overall goals were listed under each category and strategies for achieving those goals suggested. In these strategies, we incorporated Stanford-specific concerns.

The Need for a Cultural Shift and the Importance of Targets

The fact that sustainability in buildings is such a broad and varied issue that cannot be clearly defined or measured presents unusual challenges. Such a policy permeates all aspects of the building design process, and a cultural shift is required for long-term success. Recognizing this, a Land and Buildings staff member, Ted Giesing, was appointed as a half-time sustainability coordinator to help in the transition to more sustainable buildings.

One important step toward integrating sustainability throughout the entire design process was the development of sustainability checklists at each project phase to parallel the phases of Stanford's Project Delivery Process (PDP). These checklists are intended to be integrated directly into the next volume of the PDP. Theoretically, this will ensure that sustainability concerns are at least addressed throughout the building process and are introduced early in the project so that there is less chance of conflict with other construction priorities, such as program or cost.

Even so, each green building is different, and a single formula for success cannot be given; decisions of which sustainability strategies to include in a building must be made case by case. Somehow we needed a way to gauge the success of the *Guidelines* across the campus. Some of us in the working group believed strongly in the importance of targets to guide Stanford's way in achieving sustainability in its buildings. Many of us were also concerned that Stanford's best effort toward sustainability without benchmarks could end up being no different from how Stanford currently designs its buildings.

We brainstormed ways to establish an infrastructure for assessing success in incorporating green building features. At one point, we proposed a credit system modeled after LEED: instead of a universal standard for all campus buildings, we proposed creating a different set of credits for different building types (labs, residences, office/classrooms, and so forth). When this idea was rejected by the administrators, the working group figured that Stanford could not argue with bettering its own performance.

For this reason, we determined that it is important to establish performance indicators (such as energy and water consumption) within each sustainability category for each building type. Over time, we will be able to see if the Stanford building stock is improving in sustainability. In addition, the building design teams can use the range of performance data for Stanford buildings as design tools in setting targets for the performance of their new buildings. Design teams can also refer to the high-ranking buildings for visual examples of possible strategies to use. Although specific indicators have yet to be developed, the *Guidelines* were to include appendixes of sample performance indicator categories and charts (with no numbers).

A Near Failure: An Interruption from Greenpeace

As finishing touches on the *Guidelines* were being made in January 2001 in an administrative meeting, the vice provost announced that he was cutting out the appendix containing the sample indicators. He threatened to pull the plug on the *Guidelines* altogether if anyone was insistent on including them. He cited two main reasons for his decision: the recent economic downturn and subsequent pressure on the budget and the Greenpeace solar initiative.

Two months earlier, in November 2001, two Green Corps organizers working with Greenpeace came to Stanford for two weeks to start a student campaign for solar energy and green buildings. They had not researched the current happenings on campus and were unaware of SSS and the *Guidelines* development. At the end of the two weeks, the Green Corps representatives held a "solar rally," attended by about twenty students. This campaign ultimately resulted in the passage of an Associated Students of Stanford University (ASSU) resolution calling for all new university buildings to be built green and for 25 percent of Stanford's power to come from solar energy by 2010 and 50 percent by 2020.

SSS never endorsed the Greenpeace campaign or the ASSU resolution. I, among others in SSS, openly voiced my concerns about this campaign being unrealistic and lacking credibility. I was concerned that a separate student campaign toward similar goals would prove divisive and give the appearance of a disjointed effort on the part of students. I hoped that the administration would take little notice, and when nothing happened in the weeks that followed, my anxiety eased.

But in January, the vice provost had received a stack of letters demanding solar from students and the Green Corps representatives. He automatically thought that I, as a student, was in league with them and claimed a violation of trust on my part. For the first time, I was tempted to quit and remove my name from any association with the *Guidelines*. Conferring with other SSS members, I instead decided to meet with the vice provost and see how he responded to my concerns. I had already negotiated and acceded to many of his demands, but this was one point I was not willing to give up. The indicators were essential to the *Guidelines* and measuring performance.

I scheduled a private meeting with the vice provost, who immediately apologized and admitted that he could not expect me to monitor and control the actions of all students. In explaining his decision to cut the indicator appendixes, he told me he did not want to raise expectations that Stanford would improve the sustainability performance of its current buildings. I could understand that he did not want to make promises, or anything that could be perceived as one, that he felt he might not be able to fulfill, especially under tighter budget constraints of the economic downturn. I agreed to take the appendixes out of the document, but only as long as the indicators did not disappear as well. In return, he promised to include a few paragraphs about the indicators (including a pledge that the performance indicator database would be developed within a year).

I believe that in this meeting, the *Guidelines* were saved from failure. Although I believe that Greenpeace has been successful in promoting environmental consciousness in many of their various campaigns across the world, this experience with their Green Corps representatives at Stanford raised several concerns in my mind. I believe they had the best intentions but did not think through their plan very well or consider the ramifications of their actions.

Those of us working for environmental issues must work together and be educated about the causes we advocate. I consider myself a staunch environmentalist and support renewable energy, but it is only one small piece in the larger sustainability puzzle. Successful environmental campaigns require time and effort to understand where people are coming from and the complexity of local constraints. Perhaps most important, they require follow-through. Although the Greenpeace campaign might have been useful in propelling an apathetic student community into action, the manner in which it was executed at Stanford proved to be divisive and disruptive. In two weeks, it nearly ruined a year and a half's progress.

Frustrations with the Administration

The Greenpeace episode was only one of several examples of the tension with the administration. Earlier, during the development of the *Guidelines*, there were times when I felt the administration was trying to control the process too much. But I respected their judgment and could understand that they wanted to present a finished document to the public. One time, however, I was asked not to present a proposal from the working group to the ESMG until the senior administration had time to approve it.

From a student's perspective, this was one of the most important moments for me. All along, I had felt a power differential between administrators and students. The administrators called the shots and had the power to pull the plug. The fact that I had never overseen a building project or worked on producing guidelines before fed my insecurity. It was easy to be intimidated. But I knew I had a job to do: to represent SSS. So I went ahead and presented my proposal at the meeting anyway. A healthy dose of disobedience is necessary, I have found. I knew it was my right to present an idea. From that moment on, I felt ownership of the *Guidelines* project and knew it was my responsibility to see it through.

Reflections on Committee Representation

One of the most important lessons I learned from working with the ESMG is the need for diverse membership on such a committee. Diverse representation allows greater creativity and ambition in tackling a problem as tough as this one. Different forms of knowledge can be brought to the table. As I believe is the case in many other institutions, the status quo is much easier to follow than shaking things up a bit. A new pair of eyes seeing the system for the first time can add a fresh perspective for new paths that are still amenable to the old administration.

The ESMG membership was not as balanced as it could have been. Land and Buildings staff dominated the committee membership, so at times it felt as if other views were outnumbered. Even those supportive of progressive policy are subject to the demands and decisions of the senior administration and cannot push beyond a certain point. I felt that I was constantly fighting to keep the guidelines forward looking to prevent them from being watered down.

The Challenges Ahead

Now that the *Guidelines* are out (available on-line at http://cpm.stanford.edu/pdp.html), people ask me if I am satisfied with the product. I say yes and no—yes, because this is the furthest step that Stanford has ever taken with regard to green building, and no, because this should be far from the last step. There is a lot more that I would have liked to see at this point. For instance, as many people have criticized, the lack of performance targets is a large weakness. In order to tackle a problem as broad and potentially vague as sustainability, concrete goals must be set in order for step-by-step improvements to be made. But in the end, it is a question of choosing between pushing too hard for commitments and losing the battle altogether, or taking what we can get and continuing to ask for more. That is the art of negotiation: you win some, and you lose some. Will green building become common practice at Stanford any time soon? Possibly. The real test will be to evaluate how Stanford builds its buildings in a few months, or most certainly by the time this book is published. Does sustainability now sit at the table of competing priorities, as promised by the *Guidelines*?

I think the *Guidelines* hold promise, but I was often frustrated with the attitude the administrators seemed to take. Although some Land and Buildings staff were earnest about promoting green building and structuring a successful program, the general stance of the administration has been, and remains, one of skepticism and hesitation. That seems to be inconsistent with a world-class institution renowned for its cutting-edge research in a number of fields. A cultural shift, perhaps broader than we initially thought, still needs to happen.

A truly successful green building program at Stanford will require support from the top levels of Stanford administration: the board of trustees, president, and provost. They are the ones who must provide the guiding vision for green buildings and broader sustainability. Participation and commitment must come from levels above Land and Buildings. Since the vice provost of Land and Buildings is subject to close budget scrutiny, it is somewhat understandable that he is hesitant to adopt a new way of doing things that may result in additional costs. Anyone in his position cannot risk taking a long-term perspective with uncertain results (or else lose his job) that will certainly outlast his tenure. The future could hold several benefits for the university, however, including cost savings, industry and academic leadership, and positive public relations following the foothills battle, which scarred the university's environmental reputation (justifiably or not). Those, at the highest levels of administration are the only ones with the broad scope of vision and power to ensure that Stanford takes advantage of the most that green buildings have to offer.

Next Steps and Implementation

The daunting challenge of implementing the *Guidelines* still lies ahead. The ESMG met for the last time in December 2001. However, since the group's founding mission was twofold—to develop *and* integrate the *Guidelines* into Stanford's building process—I insisted on establishing an implementation group, spearheaded by the sustainability coordinator and consisting of approximately the same members of the dedicated working group.

The implementation group met for the first time (and, I hope, not the last) in May 2002 and set out plans for implementation of the *Guidelines*. In particular, the group is to focus on the creation of a database of performance indicators (the *Guidelines* promise this "during the first year of implementation"). Other tasks for the implementation group include integrating the sustainability checklists into the project delivery process, increasing publicity of the *Guidelines* and other positive environmental actions, promoting sustainability analysis of existing Stanford buildings, holding workshops for project managers, and running a green building speakers' series.

Designating the implementation of the *Guidelines* as a priority and allotting funding to support their implementation remain two of the biggest challenges that sustainable buildings at Stanford face. Staff turnover within Land and Buildings has lent a bit of uncertainty to the adoption and ultimate fate of the *Guidelines*. Money concerns were also an issue during the development of the *Guidelines*; none of the staff members working on the project had an account to which to bill their hours. I hope these problems will be solved in the future, but the fact remains that without a firm commitment from the administration and a sustainability budget to make the sustainability coordinator more than just a name, the coordinator is extremely limited in what he can do. Although Stanford's lack of money argument is to some extent understandable, all institutions, including Stanford, are required to make decisions everyday about how to use their limited resources. Relying too heavily on this reasoning can be either a decision or an excuse not to act.

Because financial and staff resources are not available within Land and Buildings, the implementation group plans to investigate a collaboration with academic departments and faculty who may be willing to give independent study units to student interns, who will help perform the work to develop the indicator databases. This is an extremely important step in bridging the gap between the academic and administrative sides of the university. Stanford can be a leader in both green building research and application.

Although ideas for a "green fund" to help fund green buildings are introduced in the *Guidelines*, no progress has been made in this area. It is unclear what it should fund (extra design fees, higher up-front costs of energy-efficiency measures, education, or green building strategies with noneconomic returns to the university, such as sustainable wood or recycled materials) and where the money will come from. SSS continues to place the development of a green fund as a top priority for the future.

Students for a Sustainable Stanford Today

SSS is now a coalition of graduates and undergraduates from diverse academic disciplines who work to educate the Stanford community about the principles of sustainability and bring about local change through this education. The group has expanded to serve as an umbrella organization for students interested in green building, green business, the general interplay between sustainable environmental practices and sustainable economics, and moving these interests into reality on campus through a practical and professional approach. While continuing work on the *Guidelines* implementation, SSS hopes to build on the experience and extend their campaign to other areas that will make Stanford a more sustainable campus. Additionally, SSS coordinates Stanford's Graduate Pledge of Social and Environmental Responsibility <http://sustainability.stanford.edu/pledge.html>, part of a national initiative among colleges and universities across the United States.

Hope for the Future

Despite the many frustrations that I have experienced over the past two years and the hopes that I have for Stanford beyond the *Guidelines*, the fact remains that the *Guidelines* are an important first step for Stanford. Stanford must be congratulated for this, and those of us who are concerned about green building can continue to work for additional changes bit by bit. Students must keep the pressure on the administration and not let them get by with anything less than promised (and keep pushing for more). In the end, no quest for sustainability will succeed if the movement itself is not sustainable. While a revolution might have been more satisfactory, I have learned to accept an evolution over time.

Notes

1. "Stanford Facts, 2002," available at <http://www.stanford.edu/home/stanford/facts> (accessed August 19, 2002).
2. Scott Gould, August 2002, personal communication.

IV

Engaging Communities, Engaging Students

10

Maintaining a College-Community Ecotourism Project: Faculty Initiative, Institutional Vision, Student Participation, and Community Partnerships

Richard D. Bowden and Eric Pallant

Allegheny College is a private, Methodist-affiliated college of the liberal arts and sciences in Meadville, Pennsylvania. Surrounded by rolling hills, small towns, and farming communities and located in the northwest corner of Pennsylvania, the college currently enrolls 2,000 undergraduates.

If we offered you an all-expenses-paid vacation to a location of your choice, where would you go? The stunning mountain ranges of the American West? A tropical paradise with white sand beaches and waving palms? An exotic foreign city filled with cultural wonders? Northwest Pennsylvania? *Where*?! As we expected. We have yet to meet someone who will choose our region as her or his dream destination. With pastoral forests, farms, lakes, and rivers, we are described as the "quiet Northwest," an apt term applied to the region by state senator Bob Robbins. Yet despite this tranquil moniker, Allegheny College, situated in the heart of this region, has taken a key role in helping to develop ecotourism in the area. This has been good for Allegheny, its students, and the larger community in which Allegheny is embedded.

In sharp contrast to the pleasant landscape of our region, the economic condition of northwest Pennsylvania has been suffering for decades. Allegheny College is located in the French Creek valley of Pennsylvania, a 1,200 square mile watershed whose 250,000 residents live in rural and small urban communities that bounce among economic stagnation, environmental degradation, and sprawl. As part of the Appalachian rust belt, human and economic resources in Meadville are struggling. Several large factories abandoned Meadville in the 1980s, driving unemployment over 20 percent. Though the jobless rate was controlled in the 1990s, poverty levels remain high. Nearly one in three

Meadville children lives below the poverty line; countywide, that number is one in five. Sprawl has created competition for struggling downtown businesses, and the jobs provided by Wal-Mart, AutoZone, and chain restaurants are generally in the low-skill, low-wage service sector. Family-owned dairy farms have been in steady decline for nearly a century throughout northwest Pennsylvania.

Environmental conditions in the region are mixed. Forestland is increasing as dairy farms are continually abandoned; however, sound forest management practices are not routinely practiced. Some farmers are adopting best management techniques to protect soil and water resources, but farm-related erosion and sedimentation continue to be major water quality issues. Ozone alerts in summer are not uncommon, and our region sits under the bull's-eye of the most acidic precipitation in the country.

Allegheny College occupies a challenging position in Meadville. As a private liberal arts college, it is an educational institution beyond the financial means of many local citizens. Lacking a graduate program or evening or weekend classes, it does not attract part-time or adult students. Despite a good relationship between the college and community, the institution is not generally considered readily accessible to area residents. Until recently, it has not been perceived as an institution that has sought to invest intellectual and practical energy in the welfare of the region.

In the early 1990s, a number of Allegheny faculty members, particularly in the Department of Environmental Science, began examining issues of local sustainability within a variety of classes. Early efforts by faculty and students in some classes examined recycling and energy consumption on campus. Gradually, however, there was a move off campus toward sustainability projects in the local community, examining issues such as land use, forestry and wildlife, energy, and environmental education. Sustainability projects were becoming collaborative efforts among faculty, students, and community members. Nine faculty members from six departments became interested in using the regional community as a hands-on laboratory for teaching sustainability within the classroom. Rather than simply having students observe what was or was not being done, there was a desire among faculty that faculty members and students together would actively collaborate with citizens, community leaders, businesses, and industry on projects that promote sustainability. Students were also increasingly interested in knowing how classroom principles could be applied to real-world situations; they sought hands-on experiences in sustainability projects analogous to those projects often conducted within natural science courses.

These movements arose naturally from the Department of Environmental Science, comprising natural and social scientists, with professional expertise ranging from political science and ecological economics to forest and aquatic ecology. All faculty have wide interdisciplinary interests and collaborate often on projects and teaching. The department has a long history of faculty research, routinely involving students as assistants for both natural and social science investigations during academic and summer months. In addition, student internships with local organizations have been popular and well supported by students for decades.

To address these growing faculty and student desires, the Allegheny College administration responded in 1997 to initiatives by the Department of Environmental Science to create the Center for Economic and Environmental Development (CEED; <http.//ceed.allegheny.edu>) to examine regional problems and provide vision that would help lead the region from economic decline toward a sustainable economic and environmental future (Pallant 2002). CEED involves faculty and students in college-community partnerships that address local and regional sustainability issues. Students are engaged in hands-on, civically engaged projects on a number of interconnected fronts: watershed protection, educational outreach, industrial pollution prevention, community visioning, sustainable agriculture, energy, forestry, and environmental justice. Students and faculty partner with numerous regional agencies, government organizations, schools, businesses, and citizens' groups. Eric Pallant served as one of the original codirectors of CEED and is currently the CEED director. Rich Bowden has served on the CEED executive committee and directs the CEED ecotourism project.

Ecotourism as a Means to Promote Sustainability

Ecotourism became one of the initial projects embraced by CEED as faculty became aware of the potential benefits of ecotourism to augment the economy and protect natural resources. We envisioned initially, and certainly naively, that by partnering with local tourism agencies and interested tourist locations and venues, we could catalyze creation of a business devoted to leading ecotours of the natural assets and destinations of our region. Ecotourism seemed a sound choice because, as defined in July 2002 by the International Ecotourism Society (TIES) <http://www.ecotourism.org>, it is "responsible travel to natural areas that conserves the environment and sustains the well-being of local people." Ecotourism differs from conventional tourism in that it explicitly promotes education about visited areas, advances sustainable use of resources, and actively avoids degradation. In addition, it assists economic development, while respecting cultural, social, and political aspects of local people. As we looked at our assets (forests, lakes, streams, rivers, wildlife, hiking trails), potential markets (nearby Pittsburgh and Cleveland), and an already strong tourism industry ($168 million annually in our county in 1990), ecotourism represented a potential means to enhance both economic development and environmental stewardship. Tourism development was cited as a means to enhance development in our "third world" Appalachian region (Nicholls 1980) and was touted as a potentially strong means to revitalize rural economics (Long and Edgell 1997, Saeter 1998).

Challenges and Strategy to Building an Ecotourism Industry

Develop a Faculty-Student Research Team

As a small undergraduate liberal arts college of fewer than 2,000 students, we do not have graduate students who can focus for extended periods exclusively on research projects. We lack research staff or extension agents found at larger universities, and as a teaching institution where faculty members teach six courses each year, time for research and scholarship is limited. Both of us have our primary training in the natural sciences (biology, soils, ecology), and we have had much experience employing undergraduate students as assistants for natural resource research investigations. Certainly we have witnessed the enormous teaching value inherent in quality research experiences for undergraduates. What has evolved for us, however, is using students to assist with community-based projects in sustainability. In some ways, there are similarities between using students for natural science and for civically engaged sustainability efforts. In both cases, students begin low on the learning curve, and faculty must invest heavily in early training. This investment, especially on the part of faculty, cannot be understated. During the academic year, when faculty time is consumed by teaching, students are important in maintaining the progress of our projects and in transferring skills and knowledge to the next cohort of student assistants. What differs, however, is that students participating in sustainability efforts are much more involved in the community and are likely to be interacting with the public, government leaders, and businesses and industry. Student assistants are ambassadors for our goals. They need to be independent and responsible thinkers with strong communications skills. For example, our early work on cataloguing local natural resource assets was accomplished primarily by a summer student assistant. She worked very competently and independently (Bowden was out of town all summer) and gained great respect from the local organizations with which she interacted. Her efforts eventually led to an academic internship with the local convention and visitors bureau. Our experience thus far is that most students act professionally and responsibly; rarely have we been severely disappointed or embarrassed.

At Allegheny, all academic departments have a required seminar that occurs in the student's junior year. In the Department of Environmental Science, this seminar often involves students in collaborative, class-wide research projects that demonstrate how to plan and execute an applied practical project. It allows students to "reach beyond the classroom" (Lempert 1996) and integrate book knowledge with practical experience and service-learning. In this class, faculty members select the topical area of study and a specific project to be addressed by the class. It becomes the students' responsibility, under faculty guidance, to design and implement the project. A number of faculty have used this seminar requirement to further CEED's goals of contributing to regional economic and environmental sustainability.

Understand Our Region and Build Community

Despite our growing knowledge of ecotourism and the great wealth of knowledge held by local officials involved in tourism promotion, local understanding of ecotourism was limited or nonexistent. Most officials were well aware of the natural assets that draw visitors to the region, but larger principles inherent within ecotourism, such as sustainable resource use, were not part of the local lexicon. Most people involved in local tourism were focused on stimulating tourism; no one was thinking of connecting tourism, natural resource protection, and cultural viability in the ways that we were. Further, many residents and business owners in northwestern Pennsylvania are opposed to environmental regulation, and environmental initiatives are often viewed with considerable skepticism. The director of a local environmental group, for example, labels himself as a "conservationist" when interfacing with the public, thus avoiding distracting controversies associated with the term *environmentalist*.

We also recognized that a local approach to ecotourism required rethinking local tourism promotion. We learned that many state efforts at promoting tourism frequently worked to favor county-specific projects and essentially, if unintentionally, promoted competition among counties. Eight to ten counties constitute the northwest corner of the commonwealth of Pennsylvania. This fragmented approach to tourism promotion was a roadblock because county-specific promotional efforts would not benefit tourists who are interested in regional tourism opportunities. Whereas we in an academic community are accustomed to thinking systematically, taking interdisciplinary approaches to problem solving, and encouraging our students to do likewise, it was apparent that the real world did not work in such an integrated manner. In an area struggling for economic survival, most local efforts focused on immediate, short-term initiatives. Focusing on long-term cooperative approaches represented for many a new view of the issue.

Local participation was key to gaining support from the community, fostering better planning, and legitimizing the decision-making process (Drake 1991; Wells, Brandon, and Hannah 1992). We believed that undertaking a civic approach to our work would nurture the interdependence of the college and community (Barber and Battistoni 1993) rather than setting up a top-down approach that could lead to imbalances of power and equity. Our movement in this direction was driven by the congruence of student satisfaction and community appreciation. Students were increasingly satisfied that they were applying classroom principles to local, practical issues that were immediately relevant. And community partners were increasingly appreciative of student efforts and community-college collaborations that helped them address current issues. Nevertheless, the desire to establish an ecotourism enterprise was driven more by faculty interest than it was by a jointly derived academic-community partnership. We knew, however, that this project would work only if there were community support. Thus, we identified people and organizations most interested in promoting ecotourism. We also became increasingly connected with members of the business community, especially those associated with tourism. Contacts included members of the local chamber of commerce, our convention and visitors bureau, and local businesses involved in travel and tourism. A travel tour company and a canoe retail and rental firm were strongly interested in our efforts as well.

Develop a Local Integrated Approach to Ecotourism

When we began our work, the ecotourism industry was primarily international in nature; little emphasis was given to domestic application of ecotourism. Wealthy North Americans or Europeans are enticed to beautiful natural areas, typically in less developed regions of the world. Web site search engines produced numerous hits for non-U.S.-based ecotourism locations. Published travel guides extolled the wonders of nature-based tourism in exotic locations; there is also a rapidly growing list of publications that explicitly feature ecotourism at international sites. During the early course of our efforts, even the flagship organization of the ecotourism industry changed its name from The Ecotourism Society to the International Ecotourism Society. Despite this international information, we found relatively little documentation on U.S.-based ecotourism, and only recently have numbers of domestic ecotourism accounts begun to grow. Thus, the two of us found ourselves without domestic models as we formulated a plan to attract North Americans to a quiet North American tourism location. The challenge was to envision ways in which we could package ecotourism locations in our region in arrangements that would be appealing to regional tourists.

Build a Model Ecotourism Product

We needed a product to pitch—something that would show how environmentally viable ecotourism might happen in this area. That product was to be a series of carefully planned ecotours that would show a prospective entrepreneur or existing travel agency how such an ecotour would look for our region. To accomplish this, two consecutive seminars for juniors constructed seven different ecotours (table 10.1). We immersed students in literature that explored the background of ecotourism, its potential benefits, and possible pitfalls. We also visited locations that might serve well as local ecotourism sites, and we met with individuals to learn how they viewed these potential opportunities.

Ecotours were constructed with sufficient detail and background information so that tour companies could efficiently organize their tours. We also provided background information for step-on tour guides (knowledgeable local guides who board buses for the duration of the tour), as well as promotional brochures for potential customers. Near the end of each ecotourism course, we invited tourism professionals to listen to student presentations of each of the tours.

Table 10.1
Ecotours constructed to show potential for ecotourism in northwest Pennsylvania

Tour	Objective	Sites included
Art and Wildlife	Examine wildlife in nature and in artistic presentation	Art gallery, art studio, national wildlife refuge
History and Nature; Grandparents and Grandkids	Show importance of natural resources in history	Fish hatchery, Oil museum, deer park
Paddle and Pedal	Explore water and watershed of biologically-rich stream	Ecotour business, historic inn, French Creek
Floating and Boating	Explore lakes and streams	Historic inn, French Creek, natural and man-made lake
Adventures in Agriculture	Show variety of farming practices in region	Dairy farm, art studio, buffalo farm, market house, organic/low-input farms
Alternative Agriculture	Focus on alternative and sustainable agriculture	Reduced-till dairy farm, buffalo farm, maple syrup farm
Fin, Fur, and Feathers	Feature wildlife, agriculture, and art for senior citizens	Wildlife museum, fish hatchery

Initial Outcomes

Public Outreach and Support

Our first public workshop on ecotourism, in March 1998, had mixed success. It included federal and local officials, people involved in the tourism industry, interested individuals, and the media. The local press reported on it favorably, and many people and businesses expressed interest in the possibilities of ecotourism. For example, the director of the Erie National Wildlife Refuge saw this as a means to attract more visitors to our relatively seldom-visited nearby refuge. Other participants were far less supportive, including one who stated that the workshop was of no value because our efforts did little to support the county represented by that person.

Shortly after this somewhat disappointing workshop came three additional opportunities to publicize our project. First, we were invited to discuss our efforts at a Pennsylvania House of Representatives committee meeting on tourism being conducted by one of our local representatives. We were a bit surprised because we were not aware that our efforts were known outside a relatively small circle of collaborators. Second, we were invited to present our work at the annual meeting of the Allegheny Watershed Network <http://www.alleghenywatershed.org/html/awnhme.htm>. Third, a nearby business association, seeing the potential benefits of ecotourism to its region, asked us to present our initiative to its members. These three opportunities came about because our community partners had promoted our project to their partners. The committee invitation came, we think, due to our involvement with the local visitors bureau, the watershed presentation came as a result of our partnership with a local conservation organization, and the business presentation was facilitated by cooperation with a senior citizens' center.

Potential Ecotours: Our Prototype for Local Ecotourism

We had some success with one prototype ecotour, entitled "Fin, Fur, and Feathers," that was designed to be used with local senior citizens; seniors have been identified as one of the fastest-growing sectors of the tourism market. This one-day, six-hour tour, with Bowden and one student acting as tour guides, took seniors to a fish hatchery, a wildlife museum, and a dairy farm that practices soil conservation to examine linkages among natural resources, conservation, and environmental issues of the region.

Our one test, which had overwhelmingly positive customer surveys, indicated that such tours could be run successfully and could attract enough clients to be profitable. Importantly to us, students put into practice some of the theoretical aspects of ecotourism, linking classroom knowledge with a hands-on community activity. Our other tours languished, however. Although we received useful suggestions from professionals who attended our class presentations and many expressed optimism that the tours would be well received, no one was stepping forward to adopt our ecotours or to facilitate movement of our tours into a business. We had been sophomoric to think that a good idea would be adopted readily into the business market.

We decided to look more closely at the economic prospects of ecotourism in our area. We conducted a market analysis and built a business plan, and were encouraged when the executive director of the local chamber of commerce informed us that the chamber receives two or three inquiries per week from entrepreneurs interested in considering Meadville as a location for starting new businesses. A student who was working on his senior thesis produced the market analysis, and a summer research assistant produced the business plan for an ecotourism enterprise.

Neither of these endeavors was very successful. Rich's academic training in forest ecology and biogeochemistry and Eric's training in soil science did not help us find business entrepreneurs. We did not know what questions to ask or really how to proceed with a sound market analysis or business plan. Analogously, CEED had once developed a business plan to create a hydroponics-aquaculture plant (Pallant 1999). That project faced the same barriers as the ecotourism business plan: lack of experience on our part and the absence of venture capital or an interested entrepreneur. And like the ecotourism idea, it languished until an unexpected call arrived from operators of a regional coal-fired power plant who were interested in developing an aquaponics facility and funded one of our interns to assist with planning the potential enterprise.

Development of an Ecotourism Web Page

At this point, our ecotourism effort had lost considerable momentum. Although we had prepared a number of tours, gathered enthusiastic support from a number of sectors in the region, and had learned that there was support and probably a market for ecotourism in northwest Pennsylvania, what we still had, primarily, was an academic study. Students had worked hard and learned, but we saw no way to move our idea into the marketplace. We thought seriously about giving up the project, but we also had a hard time letting go of all our efforts and community investments.

It occurred to us that although we were unable to give birth to an ecotourism business, we could use our tremendous database of local assets to build an ecotourism Web page. This Web site could attract those already interested in nature and might expand the ecotourism goals of environmental protection and economic development by wooing new tourists. So, again hiring a summer student and partnering more closely with the Crawford County Convention and Visitors Bureau, we catalogued and organized our vast information, begun by our first summer ecotourism student assistant, into an extensive ecotourism web page (<http://naturetoursim.allegheny.edu/>). Several things began falling into place. Rather than having an internal, wieldy, and inaccessible three-ring binder of local tour stops, we began to prepare a publicly accessible Web page for the ecotourism niche. At about this time as well, the executive director of the visitors bureau informed us of a grant opportunity to promote ecotourism in the region. The director needed funds to publish a much-needed county brochure, and Allegheny College could use funds to hire students and pay for expenses related to construction of the Web page. The new director had little experience in grant writing, but we had written many successful proposals and were happy to partner on this project. Ultimately, the bureau received a $15,000 grant from the Pennsylvania Department of Community and Economic Development to assist with ecotourism efforts.

Following two years of construction, the page was launched in May 2001, and we hosted a formal opening ceremony on campus in October, attended by college personnel, local officials, state representatives, and local media. The site was receiving over 250 hits each month by November 2002.

Just when it looked as if we would retire from our role in promoting ecotourism, an Allegheny student, for her senior thesis, proposed to develop a number of regional automobile driving ecotours for inclusion on the Web site. If we could not enable production of commercially viable bus ecotours, we could at least provide information on natural assets and environmental issues of northwest Pennsylvania for use by individuals and small groups. We must admit that we did, and still do, have some moral angst regarding the driving ecotours, which foster environmental impacts of automobile use and promote individual transportation rather than the mass transit that we had envisioned with bus ecotours. Nonetheless, the student spent her senior year developing the tours, complete with site descriptions, related information on environmental topics, maps, and useful Web links. Those who have transformed student projects into professional productions or publications, however, may recognize immediately that there was a large jump from a completed student project to a professional application. Despite this student's tremendous effort in producing a sound document, it still required two summer interns working for two months with Rich Bowden to transfer the tours (final editing, checking links, formatting, Web construction) from a written document to a Web-ready product.

Tourism Film: An Unforeseen Opportunity

Developing our relationship with the visitors bureau had unforeseen benefits. In 2001, the bureau director contacted us with an enticing offer to produce a tourism film to promote our region. Though this was not meant specifically to be an ecotourism film, the film highlights natural assets of the region. From this production, discussions have emerged on yet another possible tourism film project with the bureau.

The Future

The Web page is operating and the driving tours are on-line. Interestingly, one of our goals, to launch an ecotour business, has been partially met. The French Creek Project has begun a canoe rental service (French Creek Ecotours, <http://frenchcreek.alleg.edu/ecotours.html>), with an ecotour emphasis. Since our efforts began with support from the French Creek Project, one element of our effort has come full circle.

We are also revisiting the bus ecotours. With the extensive information prepared for our Web-based driving tours, two summer interns prepared bus ecotours that can be promoted by the convention and visitors bureau. In our region, many tourism groups attend conventions to promote excursions by tour bus companies.

We still hope that a full-fledged ecotour company will succeed here. Even if that does not occur, we now have Web site and driving tours available to ecotourists. And we will see how well our bus tours are accepted as we work with the visitors bureau to market them to bus tour agencies.

Lessons Learned

Beware of Hidden Flaws and Challenges

Despite the ecotourism principles laid out by the International Ecotourism Society, many nature-based and ecotourism companies engage in questionable marketing strategies that exploit local environments, fail to safeguard local citizenry, and generate fewer economic benefits than anticipated (Boo 1990, Chalker 1994, Honey 1999, Epler Wood 2002). Indeed, the travel industry itself has voiced concern over the fragile balance between tourism promotion and resource protection, and a local kayak and canoe outfitter, well aware of potential detrimental impacts of tourism, voiced concern that too much promotion and use of a nearby creek might be biologically damaging and aesthetically displeasing. Such warnings tempered our initial enthusiasm, suggesting a more cautious approach to our efforts. Knowledge of possible pitfalls was also important pedagogically. Some students had assumed, as we had, that because ecotourism offered so many possible advantages, it should be easy to implement in this region.

Maintain a Long-term Vision, But Create Short-term Goals

We were too ambitious at the start, hoping, and perhaps assuming, that we would spur creation of an ecotourism business in a relatively short period. When our early progress slowed and an ecotourism business was not imminent, students became disappointed, hurting our early morale. Our initial ambition and only modest successes suggested to some students that we did not know what we were doing (partially true).

We also learned that students have a short time frame for evaluating project success. Though the progress we have made in three years is rapid from an institutional standpoint, it can seem like an eternity for students. Just as others are discovering the importance of providing optimism when teaching the complexity of environmental issues in the classroom (Maniates 2002), we have learned firsthand that large-scale projects need success-bearing short-term goals that help maintain optimism and morale.

Adopt Alternative Plans

As pointed out recently (Simmons 2001), research scientists (which also includes us) often write scientific papers in a format that purports to follow a logical mode of inquiry: hypothesis, methods, results, conclusion. However, our discoveries and advances were often more serendipitous, freethinking, and creative than such writing suggests. Detours are a part of the game. Our initial attempt, for example, to document our ecotourism efforts in a chronological, historical narrative revealed how serpentine our own path has been. When initial ecotours were not adopted and no ecotourism business was forthcoming, we sought an alternative path. Building our Web site was not part of our initial plan, but it was an excellent alternative when we could not meet original objectives. Similarly, although the tourism film had not been anticipated, it was an opportunity to maintain momentum and strengthen relationships to the community.

Expect Delays in Transforming Student Projects into Professional Output

Student efforts rarely proceed smoothly. Numerous time demands often preclude students from continuing their commitments to the projects beyond their initial involvement. For example, our Web page was 90 percent constructed in the first summer, but the remaining 10 percent took an additional nine months. Producing quality products also takes time. As we partner with community members, professional outcomes are expected, and such quality is not often achieved within the confines of a given course or student experience. For example, our student thesis on driving tours was a valuable educational experience and produced most of the needed material, but transforming the thesis from an educational product to a professional Web page required a substantial commitment of additional time. Frequently, it is the charge of a faculty member to complete efforts begun by students; such completion is critical if the community is relying on efforts by the college.

Beware of Growing Expectations

As projects are completed and successes receive public recognition, both the community and the college may expect faculty members to become resources and to commit time that they might not have available. Failure to provide additional effort can lead to disappointment by students and the community.

Students also have growing expectations. Many students have come to expect that each of our required junior seminars will be project based and will be on exciting topics, and that results of their efforts will be imminent, completely successful, and highly publicized. However, not all faculty members are skilled or willing to engage in courses that are "messy" and fluid and where course outcomes cannot be ensured. Further, not all topics are of equal interest to all students. For example, one science student became captured by the Art and Wildlife tour, whereas another science student in the same class was dissatisfied by the topic of the course. When project-based courses are unavailable, the topics are of little interest, or projects do not proceed as planned, students can become disheartened, disinterested, or disillusioned.

Involve Partners from the Beginning

Our work requires the cooperation and enthusiasm of various community members: business owners, organizations, and government officials. Our partnership with the convention and visitors bureau, for example, helped us to understand tourism in the region, and we have been able to assist our community in one of the fastest-growing segments of the tourism market. In our early efforts, we did little more than provide information and occasionally seek advice from the community. In our workshops, for example, we simply invited local tourism officials, without determining their needs. As we have gained experience, we have begun evolving toward higher and more desired levels of community partnerships, wherein community members are joint decision makers and initiate action based on results of our collaboration (Paul 1987). That the visitors bureau had asked us to partner with them on a grant and approached us with the opportunity to produce a tourism film represents the kind of interactive college-community partnership that we seek. We received a very satisfying compliment recently when the bureau director told Rich Bowden that she "had really learned about partnerships by working with [us] on this project." We had thought that *we* had learned a great deal about building relationships and partnering with the community. It was satisfying to know that the learning was occurring in two directions.

Acknowledgments

We thank Juanita Hampton, executive director of the Crawford County Convention and Visitors Bureau, for helpful guidance and support; the French Creek Project for support of this initiative; Cindee Giffen, Curt Stumpf, Laura Paich, Kelley Angleberger, Maureen Copeland, and Chris Shaffer for research assistance; Bud and Carol Luce for their encouragement; and Tim Sipe for editorial suggestions. This project has been supported by the Heinz Family Endowments, the Pennsylvania Department of Community and Economic Development, and a sabbatical grant from Allegheny College. This publication is a contribution to the Allegheny College Center for Economic and Environmental Development.

References

Barber, Benjamin R., and Richard Battistoni. 1993. "A Season of Service: Introducing Service-Learning into the Liberal Arts Curriculum." *PS: Political Science and Politics* 16:235–240.

Boo, Elizabeth. 1990. *Ecotourism: The Potentials and Pitfalls.* Vol. 1. Washington, D.C.: World Wildlife Fund.

Chalker, B. 1994. "Ecotourism: On the Trail of Destruction or Sustainability? A Minister's View." In *Ecotourism: A Sustainable Option?* (pp. 87–94). Edited by Ertlet Cater and Gwen Lowman. New York: Wiley.

Drake, Susan P. 1991. "Local Participation in Ecotourism Projects." In *Nature Tourism: Managing for the Environment* (pp. 132–163). Edited by Tensie Whelan. Washington, D.C.: Island Press.

Epler Wood, Megan. 2002. *Ecotourism: Principles, Practices and Policies for Sustainability.* Burlington, V: International Ecotourism Society.

Honey, Martha. 1999. *Ecotourism and Sustainable Development: Who Owns Paradise?* Washington, D.C.: Island Press.

Lempert, David H. 1996. *Escape from the Ivory Tower.* San Francisco: Jossey-Bass.

Long, Patrick, and David L Edgell. 1977. "Rural Tourism in the United States: The Peak to Peak Scenic Byway and KOA." In *The Business of Rural Tourism:*

International Perspectives (pp. 61–76). Edited by Stephen. J. Page and Don Getz. London: International Thompson Press.

Maniates, Michael. 2002. "Of Knowledge and Power." In *Encountering Global Environmental Politics: Teaching, Learning, and Empowering Knowledge.* Edited by Michael Maniates. Lanham, Md.: Rowman & Littlefield.

Nicholls, L. L. 1980. "Regional Tourism Development in 'Third World America': A Proposed Model to Appalachia." In *Tourism Planning and Development Issues* (pp. 283–294). Edited by D. E. Hawkins, E. L. Shafter, and J. M. Rovelstad. Washington, D.C.: George Washington University.

Pallant, Eric. 1999. "Raising Fish and Tomatoes to Save the Rustbelt." In *Acting Locally: Concepts and Models for Service-Learning in Environmental Studies* (pp. 89–98). Edited by Harold Ward. Washington, D.C.: American Association for Higher Education.

Pallant, Eric. 2002. "Allegheny College: Bringing Sustainability to Northwest Pennsylvania." In *Teaching Sustainability: Towards Curriculum Greening* (pp. 405–414). Edited by Walter Leal Filho. Bern, Germany: Peter Lang Scientific Publishers.

Paul, Samuel. 1987. "Community Participation in Development Projects: The World Bank Experience." *Discussion Paper No. 6.* Washington, D.C.: World Bank.

Saeter, Jens Aarsand. 1998. "The Significance of Tourism and Economic Development in Rural Areas: A Norwegian Case Study." In *Tourism and Recreation: Rural Areas* (pp. 235–245). Edited by R. Butlema and J. Jenkins. Chichester: Wiley.

Simmons, Robert. 2001. "Sense and Sensibility: Could Literature Teach Us How to Release Scientific Writing from Its Straitjacket?" *Nature* 411:243.

Wells, Michael, and Katrina Brandon, with Lee Hannah. 1992. *People and Parks: Linking Protected Area Management with Local Communities.* Washington, D.C.: World Bank.

11

Teaching for Change: The Leadership in Environmental Education Partnership

Paul Faulstich

Pitzer College is a member of the Claremont Colleges consortium and is located in Claremont, California, about thirty-five miles east of Los Angeles. Within Claremont, Pitzer's educational philosophy is singular; Pitzer strives to enhance individual growth while at the same time building community. A private, liberal arts institution, Pitzer enrolls about 900 students, and the campus is adjacent to the Bernard Biological Field Station, featured in this chapter.

Humans are transforming earth's landscape from a natural matrix with pockets of civilization to just the opposite. Most of us realize that this pattern is not sustainable. I live and work in Claremont, California, a charming college town in the midst of suburban sprawl. The town has a central village of terminally tasteful, overpriced bungalows nestled in the shade of tall, largely exotic trees. Indeed, most of the landscape of this "city of trees and Ph.D.s" has been imported; only a remnant parcel of coastal sage scrub that the Claremont Colleges have reluctantly preserved remains. The coastal sage scrub ecosystem, once the prevalent indigenous plant community in the Claremont region, is now endangered as a result of sprawl and inappropriate development. It was partly our experience of this disjunction between environmental past and present that led me to develop Pitzer College's Leadership in Environmental Education Partnership (LEEP).

LEEP provides place-centered environmental education for eight- to twelve-year-old children, while training college students in principles of environmental education that prepare them for the fields of teaching, environmental advocacy, and environmental nonprofit administration. To present an overall assessment of this endeavor, I begin with a basic description of the LEEP program, followed by a discussion of its founding, development, and some of the challenges it has faced. I conclude with comments about the constant and ongoing efforts required to sustain LEEP.

The LEEP Program

Since 1996, LEEP has enabled approximately 150 college students and 870 schoolchildren from four elementary schools in the Claremont Unified School District to study ecological and environmental issues at the Bernard Biological field station. The Field Station, an 85-acre parcel contiguous with the campuses, contains an unusual variety of habitats. In addition to coastal sage scrublands, it harbors a constructed aquatic habitat (pHake Lake), a riparian zone, coastal oak woodlands, and vernal pools. It provides refuge to rich and diverse plant and animal populations, including such sensitive native plant and animal species as the Santa Ana River woolly-star, Nevin's barberry, Riverside fairy shrimp, southwestern pond turtles, coastal whiptail lizards, and cactus wrens. During an eleven-week unit, classes of school children visit the field station once a week for three hours to participate in interdisciplinary study of its native coastal sage scrub ecosystem.

LEEP provides hands-on lessons in environmental science, ecological diversity, human ecology, environmental awareness and appreciation, habitat restoration, and pollution prevention. Children and their teachers observe the habits of fauna, examine owl pellets and animal scat, study flora, gain knowledge of vernal pools, make sample collections, carry out laboratory analysis, and record their findings in field books. They participate in clinics addressing various environmental topics, including ethnobotany and local Indian traditions. They also carry out simple environmental restoration projects that improve biologically degraded portions of the station. These activities encourage the development of an environmental ethic and ecological identity. For some students, LEEP is their only connection with the beauty and diversity of our native ecosystem.

The four schools that currently participate in the program are relatively diverse, each with unique features. Mountain View School's student body is 38 percent Caucasian, with the remaining 62 percent representing other ethnicities. Vista del Valle serves a multiethnic population, and more than 68 percent of the students qualify for Chapter 1 funding. Sumner-Danbury is a joint campus where standard education students and orthopedically disabled and health-impaired students are fully integrated. Sycamore Elementary provides a multiage developmental program that serves students who speak eleven different languages. Of the 140 students who participate in LEEP each year, approximately 61 percent qualify for free or reduced lunch. Through LEEP, these children engage in cooperative problem solving and participate in activities that foster environmental responsibility and point toward sustainability. The children then go back to the classroom and connect their learning with their studies of biology, natural history, local prehistory, current events, and Native cultures.

Pitzer students in my course entitled "Theory and Practice in Environmental Education" <www.pitzer.edu/env-ed> serve as instructors for the elementary schoolchildren. In the course, college students are organized into four teams, each paired with one of the participating elementary schools. Over the course of the semester, the college student teams guide the schoolchildren's weekly visits to the field station. The children develop a rich and gratifying relationship with both the field station environment and their college mentors. Weekly, the college students meet as a class to explore larger theoretical issues related to their mentoring and to assess the progress of the children's learning experiences. Activities conducted through LEEP align with the California Content Standards for grades 4 through 6 in science, language, and history/social science. Field books, writing prompts, science exemplars, graphic assessments, and final portfolios attest to the balanced and rigorous nature of the curriculum. In addition to providing schoolchildren with much-needed environmental education, LEEP also exposes them to the college endeavor and provides them with college students as role models and mentors.

Our collective philosophy in LEEP is to approach environmental education in the spirit of celebration. We want to celebrate the land and its human and natural histories. Although we do not shy away from discussion of environmental degradation, we also do not want to fill our curriculum with examples of environmental abuse. "Environmentally correct" curricula can make children feel estranged from nature rather than coupled with it. My hope is that LEEP will help students to reinhabit our bioregion, to dwell in ways that acknowledge ecological limits and engender sustainability. By facilitating early environmental education, LEEP aims to counter alienation from nature and endow youth with a strong and lasting kinship with the earth. Imprinting is deep learning at a critical stage of development, wherein an individual attaches momentous meaning to an object separate from the self. It is part of the natural development of all animals and is not easily unlearned. Early, deep exposure to the wonders and workings of nature can facilitate such an imprinting, a lifelong respect for the environment and a commitment to conservation. For this reason, outreach to schoolchildren is an important component of our efforts toward sustainability.

The mentor teachers and principals of the partnering schools form a motivated, engaged board of advisers. In addition to their central role as facilitators of the partnership, the board oversees curriculum planning, conducts field observations, and makes recommendations for strengthening the program. The participating schools share the results of their learning with the larger community through an annual open house at the Bernard Field Station that includes a family "scavenger hunt" (e.g., "find some scat and determine what animal left it and what it ate") and a display of student journals, photographs, art, and experiments connected with the project. Community leaders, parents, and educators come together to celebrate the learning and community impact of this collaborative effort.

LEEP is the cornerstone of environmental studies for our local public schools and has inspired a number of spin-off programs. One partnering school, for example, has developed an upper-grade science rotation that correlates with LEEP. In the spring term, the school offers students the opportunity to study one concept in depth. Students select from various science classes, including earth science ("Geology Rocks!"), chemistry ("Wait! Don't Mix Those!"), and environmental studies (LEEP). Another school has implemented green groups, including a recycling initiative and a campus relandscaping program that emphasizes greater use of native plants.

Founding and Development of LEEP

The history of LEEP is a web of intertwined ideas and motivations. The program emerged in 1996–1997 independently but concurrently with several important events, including the revision of Pitzer's Social

Responsibility Guideline and the naissance of the Claremont Educational Partnership.

Founded in 1963, Pitzer College is a liberal arts institution with a curricular emphasis in the social and behavioral sciences. Enrolling some 800 students, Pitzer is part of the Claremont Colleges, a consortium of five undergraduate colleges and (now) two graduate schools. Six of the campuses are physically contiguous, and all share such facilities as a central library, bookstore, and the Bernard Biological Field Station. In keeping with its 1960s heritage, Pitzer's educational philosophy strives to enhance individual growth while building community and is associated with the promotion of progressive social change. Students create their own academic programs in close collaboration with their faculty advisers. There are no lists of requirements; rather, students are guided by a set of educational objectives, one of which articulates a commitment to "Concern with Social Responsibility and the Ethical Implications of Knowledge and Action." By undertaking social responsibility and examining the ethical implications of knowledge, students learn to evaluate the effects of actions and social policies and take responsibility for making the world we live in a better place.

At Pitzer College, social responsibility is defined as awareness, knowledge, and behavior based on a commitment to the values of equity, access and justice, civic involvement, and environmental sustainability, and it is rooted in a respect for diversity, pluralism, and freedom of expression. To improve implementation of this educational objective, Pitzer introduced a specific guideline during the 1995–96 academic year that requires students to participate in a semester-long community-based service project. Students may pursue one of several options to meet this guideline, but the preferred method is an experiential-learning placement in the context of a course (e.g., LEEP). Following the introduction of this new guideline, Pitzer began to encourage its faculty to experiment with service-learning courses and to develop experiential learning projects.

While the vast majority of social responsibility courses are driven and sustained by the research interests of individual faculty members, LEEP emerged more out of passion than expertise. I am trained as a cultural anthropologist, and I direct LEEP largely as an add-on to my other responsibilities. As an academic generalist with diverse interests, I am engaged in preparing students not only to learn but also to act effectively on their values and to participate in their communities. I strive to encourage proactive and intelligent responses to our social and ecological dilemmas. My academic strengths lay in a broad human ecology, which teaches that diversity, interdependence, and whole systems are fundamental to us and to the health of the planet. This is the passion that was the impetus for LEEP.

The introduction of Pitzer's new social responsibility guideline roughly coincided with the signing of the charter for the Claremont Educational Partnership, an arrangement between the Claremont Colleges and the Claremont Unified School District to promote increased cooperation between the colleges (individually and collectively) and the local public schools. It was formed with the conviction that a strong school system would enrich the community by fostering well-informed leaders for the next generation. Both the colleges and the school district benefit from these cooperative efforts. The public schools enjoy greater access to college-based experiences that include volunteer college student teachers, faculty development programs, the expanded use of technology in education, and greater library resources. And the colleges gain hands-on experience in the K–12 classrooms for their student teachers, interaction with potential future college students, and opportunities for students and faculty to participate in and grow through community service.

At the time of the signing of the charter for the educational partnership, a colleague of mine was director of the Pitzer Conflict Resolution Studies Program, which was already working with the public schools to implement mentoring and youth education projects. As a member of the partnership's new steering committee, she had substantive interaction with numerous local public school administrators. Through the partnership, we raised the idea of LEEP, identified appropriate schools, and made preliminary arrangements to implement the program.

With a $20,000 seed grant from Edison International, we purchased basic supplies, published a field book for student use, organized training workshops, and provided stipends to the mentor teachers. A number of foundations and organizations, including Singing for Change Charitable Foundation and the San Manuel Band of Mission Indians (the LEEP curriculum includes a component on Native American ethnobotany), have provided additional funding. The school district has provided matching funds, in-kind support, and release time for teachers. Pitzer College supports LEEP by offering "Theory and Practice in Environmental Education" as a regular part of the curriculum, providing assistance with grant writing, and maintaining the Bernard Biological Field Station.

Challenges

In the spring of 1997, the Claremont Colleges voted, amid significant controversy, to add a seventh college to the consortium: the Keck Graduate Institute of Applied Life Sciences. The colleges' board of fellows voted to give the Keck Institute, a commuter biotechnology college with strong ties to industry, 11.4 acres of the Bernard Field Station for its campus, despite overwhelming opposition by faculty and students and the existence of alternative sites. Many Claremont citizens, including representatives of the local Native American community (Gabrieliño-Tongva), for whom the land is an important cultural resource, opposed building on the field station. They gathered signatures for a ballot referendum, produced bumper stickers ("Save the Field Station: Claremont's Wild Heart"), entered floats in Claremont's annual Fourth of July parade, organized street corner demonstrations, and carried out community-wide leafleting. Ultimately, a lawsuit filed by a citizens' group, Friends of the Bernard Biological Field Station, led to an agreement to protect half of the station for a fifty-year period.

Open and ecologically sensitive land undoubtedly will become rarer in southern California, so if the field station continues to serve as a site for research and study, it will be even more valued, and the odds of its preservation will increase. All seven colleges in the Claremont Consortium contribute financially to its maintenance, and each college provides student and faculty access for study and research. LEEP currently represents the only public access. During the public debates about building on the field station, a number of community members expressed a desire for more public access. LEEP has significantly increased both academic use and public access to the property. Many people in Claremont view LEEP as a critical component in the movement to preserve natural habitat by educating future voters and policy makers to the value of this precious local resource.

Besides the obvious threat to the remaining coastal sage scrub ecosystem, the controversy over building on the field station raised other issues, particularly with regard to the role of advocacy in the LEEP program. Both college and elementary students participated in public debates about the future of the field station. Invariably, these students had been exposed to the field station through LEEP. They testified to the Claremont City Council, attended Planning Commission hearings, and published letters in regional newspapers. In 2001, Students for the Field Station, a group of college activists, staged a protest demanding the preservation of the field station in perpetuity. They barricaded entrances to the Claremont University Consortium's main administrative building by chaining themselves to barrels filled with concrete. The police used forklifts to remove the barrels and arrested fifteen students for misdemeanor criminal trespassing and willful disruption. This protest secured Pitzer the number two place on *Mother Jones* magazine's annual list of Top 10 activist campuses (September–October 2001).

Such activism has annoyed top administrators of the consortium, who have been frustrated by community efforts to preserve the field station. The elementary schoolchildren's role in the debate has sparked particular controversy. For example, children from one of the LEEP schools recently participated in a student-generated activity in which they envisioned and sketched improvements to the field station. Designed to teach students how to think, not what to think, this activity generated a number of suggestions. Proposals to plant fruit trees and remove the native poison oak indicated that this exercise was not a one-sided activist's approach to the problem. Other suggestions included installing birdhouses, planting native shrubs around the perimeter fence, and mounting signs telling people to stay on the trail. There were some references to maintaining or expanding the station, but the majority of the suggestions did not address contested land use issues.

Nevertheless, the CEO of the Claremont University Consortium responded, "While the school children's ideas may be non-confrontational, I do object to the idea that they are being taught to 'plan' for the future of the Field Station. It is a small leap from that to advocating that the property remain a field station and/or completely undeveloped in perpetuity." In another memo, the CEO noted that "such actions could put the Colleges' and school district's support [of LEEP] in serious jeopardy." I was told that we were using the children as pawns in the political battle to preserve the land. I raised objections to these responses and noted that the CEO's intention appeared to be to discredit and dismantle LEEP. I saw this as a curtailing of academic freedom and urged the consortium administration to leave curriculum planning to the professional educators engaged in the program.

Discussion of the land use controversy is decidedly not a sanctioned element of the LEEP curriculum, but the request that we avoid addressing the topic was an infringement on the process of democratic education. It is a fact that in our community, the field station is a contested parcel of land. Should college student instructors tell schoolchildren that they are not allowed to discuss this issue, even when the conversation develops naturally in the course of educational discourse? I discuss at length with my students the role of advocacy in environmental education. Although activism has an important place in many environmental education programs, LEEP is founded on the fundamental belief that children will develop their own passion to protect nature. Our role is to facilitate exposure to and knowledge of the local natural world. In the recent past, I asked my students to avoid discussion of the field station controversy, but the CEO's memo has alerted me to the inappropriateness of such a rigid position. The controversy is a central piece of the history of the land and deserved to be addressed. The line between service-learning and activism becomes blurred when students begin to love a reserve that administrators want to bulldoze.

Continuing Efforts

LEEP is a collaborative endeavor, which is exactly what makes it uniquely valuable and sometimes difficult. The complexities of collaboration are evident in the partners' differing perspectives on the relationship between environmental education and activism. The LEEP board of advisers approached the field station controversy as collaborators, acknowledging broad accountability and discussing how collectively to resolve the conflict. A representative from the board met with the school district's assistant superintendent, who agreed that the situation was of minor concern—the program had violated no educational codes. After reaffirming the value of LEEP to the school district, the assistant superintendent also stressed that we must be careful not to inhibit the flow of ideas. Ultimately, Pitzer's administration came out in strong support of

LEEP, and the president of the college noted that the envisioning activity was "a good and appropriate tool for teaching and learning."

Successful partnerships should not be measured by the absence of conflict, for this may simply be an indicator that difficult issues are not being addressed. Effective partnerships acknowledge the inevitability of conflict, and evaluation should be based on the extent to which conflict is resolved to the satisfaction of the partners. The LEEP program successfully emerged from the field station controversy, and all partners surfaced with a renewed commitment to the goals of the program. We successfully used the conflict to strengthen the collaboration. We used it as a springboard for discussions leading to consensus on issues of academic freedom and the value of outdoor environmental education. On the heels of the controversy, LEEP received a Circle of Excellence Award from the National Council for the Advancement and Support of Education. In the end, we were able to celebrate as well as critique ourselves.

LEEP is a collaborative effort, which is exactly what makes it uniquely valuable and sometimes difficult. The complexities of collaboration are evident in the differing perspectives on the relationship between environmental education and activism that the partners hold. The context and motivation for the collaboration differ somewhat between Pitzer, the partnering schools, administrators of the Claremont University Consortium, the Claremont Educational Partnership, the school district, and the students. Since disagreements arise from such differences, the partnership benefits from provisions for mediation and conflict resolution.

A sustainable future depends on teaching children to respect nature and each other. An important aspect of this educational process is helping students develop their individuality as well as their responsibility as members of the more-than-human community. Toward this end, Pitzer's College Council—the governing body of the college, which consists of faculty, student, and staff representatives—has adopted this Statement of Environmental Policy and Principles: "Pitzer College strives to incorporate socially and environmentally sound practices into the operations of the college and the education of our students. Pitzer exists within interreliant communities that are affected by personal and institutional choices, and the College is mindful of the consequences of our practices. A Pitzer education should involve not just a mastery of ideas, but a life lived accordingly. We are thus committed to principles of sustainability, and dedicated to promoting awareness and knowledge of the impacts of our actions on human and natural communities."

By engaging children in the understanding and appreciation of their local environment, LEEP aims to foster values of citizenship and social

Reflections from a LEEP Sixth-Grade Teacher

It was cold at the field station today, but the drizzle was light. Seven college students were guiding seven teams of my students over eighty-five acres of protected sage scrub. I was backing off, wandering between groups, and letting the college students do the teaching.

Tadpoles in evaporating vernal pools. Will they survive until their pools fill up again?

A forty-foot thin black line on the mud. Looking closely, we see it is a line of drying toad eggs. Thousands upon thousands. Never to hatch.

Two vultures overhead. "What kind of birds do you think they are?" A cockatiel they guessed. "What kinds of birds do you see circling in movies or cartoons?" I prompt them. "Are those the kind you see when you hit your head?" they wondered. I just told them that they were vultures.

Coyote tracks in the mud.

And then, roaming alone between groups, I saw a small clump of some type of cereus cactus in bad shape. Decomposing, I thought. But then I saw some fresh cuttings, newly chewed. Squatting down I saw what I had only thought before to be dirt. It was rabbit scat; so thick I didn't recognize it.

The remains of the fresh cactus cuttings led back into a bramble of undergrowth about 20 feet in diameter. Dropping to my belly and lying prone, I waited. Moving closer, ever so slowly, I went into the undergrowth careful not to disturb anything. Ants. Droppings. Rotting sticks. Cactus cuttings led the way. Sitting upright was a black-tailed jackrabbit. She was quiet and still and looking right at me. I could see the whiskers half way up her face; the ones above the eyes, moving.

I turned my head and in one move she was gone.

I didn't think I would see anything more beautiful that day, but I did. All seven of the groups of students had been drawn to the largest vernal pool at the field station. Thousands of tadpoles had recently emerged.

And then he held her hand.

Ben Lopez [pseudonym] cannot work in groups. He is intelligent, but only cares about his grades. He argues every day with someone. I have never seen him help anyone, except when I ask him to do so, and then he does so only reluctantly.

(*continued*)

Shy Nataly Yen [pseudonym] started to climb over a ten-foot mound of dirt to get to the other bank of the pool. Following Ben, her feet were slipping in the mud. No words were spoken, but Ben stood on top of the mound and held out his hand. Hers slipped into his. Gently, Ben helped her to the top and then let go easily when she made it.

Rabbits, deer, robins, and sometimes
People hear noises in the wild,
In the bramble, in the bushes, on
The side of winter pools of water, and
They jump, or freeze, or the hair on the back
Of their necks stands on end, and sometimes
People hear feet slipping on a mound of mud
Not fleeing, not freezing, but reaching out
Extending

Joe Tonan

responsibility. It provides college students with opportunities to teach elementary schoolchildren from diverse backgrounds about environmental concerns in the community. College students who participate in the LEEP program gain a respect for nature and habitat preservation that influences their attitudes toward the environment and, often, their choice of careers. One purpose of LEEP is to train future (and current) educators to create learning environments that offer every child the opportunity to gain a deeper understanding and appreciation of the human and natural community.

Revitalizing communities is key to ecological health and social harmony. Our current environmental crisis is symptomatic of our fractured relationship with the natural world and with each other. We are unlikely to succeed in appreciating and restoring the natural environment if we lack the knowledge and passion to restore human communities. Together with the participating elementary schools, Pitzer addresses both of these critical concerns through the LEEP program (see the box).

In many ways, LEEP is about extending: extending our learning outside the classroom, extending our relationship with the world, extending our understanding of others, extending our sense of community. In keeping with Pitzer's educational objectives, students learn to evaluate the effects of actions and social policies and to take responsibility for making the world in which we live a better place.

Acknowledgments

This chapter was completed while I was a Fulbright Senior Scholar at the Centre for Resource and Environmental Studies, Australian National University. I thank Robert N. Fabian at Emory University for his careful editing of this chapter.

12

Restoring Natural Landscapes: From Ideals to Action

Allen Franz

Marymount College, Palos Verdes, is a Catholic-affiliated two-year college in southern California, surrounded by affluent suburban neighborhoods. The college overlooks the Pacific, with urban Los Angeles twenty-two miles north.

How hard can it be to establish native plantings as a service-learning project on a college campus? It sounds like a win-win proposition: students learn about native plants and ecology, and the college saves on watering, fertilizer, and other landscape maintenance costs. Yet even with a supportive administration, there is more to it than just digging holes and inserting plants.

This chapter focuses on landscape management, and in particular on stewardship of biological resources through habitat restoration, with an aim to helping to preserve biological diversity, the most extraordinary resource on earth. At the same time, the habitat focus serves as a means to encourage responsible stewardship of other natural resources, such as water and soil, and as a way to reconnect with local landscapes and reinforce a sense of place. As I have come to appreciate, altering institutional practices requires dedication and persistence, and influencing deeper mind-sets relating to landscapes and biological resources can be even more challenging.

The Institutional Framework

Marymount College, Palos Verdes is a small (750 student) liberal arts college in the extreme south of Los Angeles County, in southern California.[1] Marymount's thirty-acre campus is situated on a scenic bluff overlooking the coastline, with Santa Catalina Island floating invitingly to the south.

From the east edge of the campus, there are views across the Los Angeles Basin to the often snow-capped, two-mile-high San Gabriel, San Jacinto, and San Bernardino Mountains. Thanks largely to the Pacific Ocean and its sea breezes, the campus environs enjoy a year-round mild climate. Yet notwithstanding Marymount's idyllic setting, it can be a challenge to advance environmentally sensitive ideas and sustainable practices.

Although southern California is by nature a semidesert region, Marymount's campus offers a luxuriant green landscape. A crew of gardeners works full time transplanting, watering, fertilizing, trimming, mowing, combating weeds and pests, removing litter and leaves, and carrying out assorted other tasks involved in maintaining the verdant greenscape of the campus "front yard," the north side, through which most visitors enter and where campus facilities are concentrated. The front yard greenscaping consists almost entirely of nonnative plants, which require intensive watering, regular application of soil amendments, and other forms of special care to enable them to survive and appear healthy and attractive. This high-maintenance landscape also requires an infrastructure of irrigation lines, mowing machinery, and other equipment. In contrast to the carefully managed front yard of the campus, the college's approximately ten-acre "backyard" is unmaintained because of the perceived low benefit in proportion to the projected costs of landscaping.

There is a clear rationale for the intensive front yard greenscaping: the scenic location and verdant grounds are among the more distinctive attractions of the college and are prominently featured in recruitment materials. For a tuition-driven private college with only minimal endowment funds, attracting and retaining tuition-paying students is critical to the institution's survival, and the location and greenscaping are important in creating a favorable impression and distinguishing the school from its competitors. Luxuriant landscaping is also good public relations with the upscale community around the college. Given that the college periodically has to petition the surrounding city for permits to hold special events or modify its facilities, it is prudent to maintain an attractive front and avoid offending neighbors.

Jim Reeves, Marymount's vice president for college operations, articulating the college's priorities in landscape management, emphasizes the desire for an atmosphere conducive to academic pursuits, safety and healthfulness, and aesthetic values. The campus is kept strikingly clean and green, and care is taken also with less visible practices; for example, in selecting fertilizers and pesticides, the college seeks nontoxic or least-toxic alternatives.

In some areas of the campus, such as buffer zones bordering neighboring residences and strips adjacent to streets and sidewalks, plant selection is dictated by city requirements. The city's list of accepted species is long on exotics and short on natives. One element of the college's current expansion plans involves incorporation of native plants in the proposed new landscaping. Total costs associated with landscape management are around $300,000 a year, or roughly 2 percent of the college's overall operating budget—far less than faculty salaries or institutional technology but nonetheless a significant expense.

Daniel Coca, a former Marymount student who returned to take a position in the college admissions office, remarked on the contrast in his perspective on the greenscaping from his student days to his current position. As a student, he appreciated the cleanliness and peacefulness of the campus but never gave it too much thought. It has been eye-opening for him to learn that the admissions office engages in detailed landscape planning in preparation for open house events for prospective students and for high school counselors who may influence students' choice of colleges. Flower beds are often planted with new stock, and watering schedules are intensified in the weeks preceding a major event, to green up the campus. Watering schedules are then cut back sharply several days before the events, so that there will be no marshy or muddy areas when the visitors arrive. Weather permitting, as it usually does in southern California, open house events are typically organized around outdoor lunch buffets that highlight the greenscaping and scenic views.

Where I Fit In

Marymount is a teaching-centered rather than research-centered college, and I am employed primarily as a classroom educator. I am primarily an anthropologist and for over twenty years a member of the teaching faculty at Marymount College, but both my personal life and my role at Marymount are also shaped to a considerable degree by environmental interests. Indeed, I see anthropology and environmental studies as necessarily interrelated. On the one hand, humans are products of their natural as well as cultural contexts, and on the other hand, human behavior is in certain respects the most significant variable affecting natural environments.

I have long believed that many of the most important lessons are learned outside the classroom. Even academic lessons are often better learned in conjunction with real-world applications rather than simply as conceptual abstractions. I have introduced a variety of field activities to reinforce classroom lessons and as co- and extracurricular activities for those not in my classes. Through participation in these projects, many students have gained both a firmer grasp of classroom concepts and a bridge between ideas and actions.

As I have gradually become rooted at Marymount, I have simultaneously become more interested and involved in the community and region in which I live and work. After years of packing up whenever possible for weekend escapes and vacations away from the metropolitan landscape of Los Angeles, I began to ask myself why I was turning my back on natural areas near where I live. Furthermore, if I value the experience of reasonably natural environments, then don't I have a responsibility to act to preserve what I value, particularly when it is threatened, as has increasingly been the case for natural open spaces in coastal southern California and elsewhere? If I also hope for others to value natural places and resources, shouldn't I contribute to conserving and restoring natural areas, so that others can experience them too? After all, how is someone who grows up in an entirely human-designed and-constructed environment ever going to experience wonder in nature and come to value it?

As a result of this progression of thoughts, I gradually set about reshaping both my personal and professional priorities. I tried to rework my personal lifestyle to become more sustainable, more environmentally sensitive, and more place based. I became increasingly involved with local organizations concerned with land use policy and practice and related environmental concerns.

On the Marymount campus, I gradually expanded my instructional curriculum to include courses in physical geography, ecology, and other fields relating to human adaptations and relationships to landscapes, and I began promoting co- and extracurricular activities that address responsible stewardship of the local environment. In a parallel development, my research interests in anthropology shifted to questions relating to how southern Californians perceive and relate to the natural world and to the political economy of natural resources and open spaces in southern California.

GAP and MOVE: Facilitating Student Environmental Action

Although the college community includes people from diverse backgrounds and religious orientations, Marymount College is rooted in a Catholic heritage, with a mission emphasizing not just the transmission of knowledge but also the search for values to shape a meaningful philosophy of life. More so than in many other schools, students and others in this college community are encouraged to contemplate their ethical foundations and to become constructively involved both on campus and in the community. The strong institutional support for service-oriented activities has facilitated the introduction of environmental activities on and off campus, including native plantings and habitat restoration.

The most environmentally oriented extracurricular program at Marymount has been an initiative called the GAP (Global Awareness Program), which was subsequently merged with another cocurricular student organization, MOVE (Marymount Opportunities in Volunteer Experience). Much of the credit for getting GAP off the ground goes to former student Eddie Sison. Subsequent student leaders, particularly Lupe Gonzalez and Eduardo Flores, helped foster GAP support for native plantings and habitat restoration work both on and off campus.

Why bother with native species? To begin with, natives have evolved in situ and are adapted to the climate and soil, and to other species living in the ecosystem. In practical terms, this means they generally do not require the substantial watering, soil amendments, or other investments of time and resources that many nonnatives do. Native plants are also the base of the food chain for almost all other native life forms. This point is particularly significant in regions like coastal southern California, where development has already displaced most native habitat and eliminated many species. Because of its complex geography, California has harbored exceptionally diverse communities, both human and biological. Paralleling its prehistoric and historic cultural diversity, California is second among the states, behind Hawaii, in the number of endemic life forms (creatures naturally occurring nowhere else in the world), and it is second only to Hawaii in the number of threatened and endangered species.

As a consequence of evolutionary adaptations, many animals have unique, highly specialized adaptations to particular plants or plant sets and have, for instance, digestive enzymes that are specialized to handle only specific host plants. In our local area, for example, we have two distinct endangered species of blue butterflies, each of which can feed on only a specific host plant during its larval stage. The host plants are declining steadily in the face of development; if the host plants disappear, ripple effects may have an impact on an array of other species as well. Preserving and restoring native plants on the peninsula, then, is not simply a nostalgic attempt to restore a romanticized sense of place, but a way to extend the lives of other native species—biological resources for the ecosystem and, potentially, for us humans.

Jim Reeves has been consistently receptive to GAP/MOVE initiatives, and in 1990 he gave GAP the green light to install some native plants along a slope of undeveloped land on the back side of the campus. We acquired a selection of native trees and shrubs and organized work parties to transplant and periodically water the new plants until they took root in their new home.

All the plants survived transplanting and grew rapidly during their first spring, but over the summer disaster struck. Because the back side of campus is not landscaped or maintained, each rainy season (winter and early spring), it produces a crop of mustard, wild oats, and other weedy annuals. These plants die and dry out over the summer, so the county fire marshal requires that they be mowed or disced (plowed under) to reduce the fire hazard. The college annually contracts with an independent business to disc the unmaintained areas, and therein lay a problem. Our plantings were arrayed along the upper margins of the unmaintained land, and the tractor operators hired to disc the land did not receive any notice that there were plants that should be spared. As a result, even though most of the plants were staked and clearly different from the weedy annuals, about half of our plantings—the smaller specimens and those farther down the slope—were disced and destroyed. Although I brought this to the attention of administrators and asked them to inform the private contractors that our plantings should be spared in the future, nonetheless the same thing happened again the next summer. Only a handful of the original plantings remained.

Having been thus frustrated by the results of efforts to extend plantings in the unmaintained areas of campus, the GAP group decided to try introducing native plants into the landscaped areas on campus. Our pilot project was the planting of four California sycamores *(Platanus racemosa)* on the lawn south of the Administration Building. The underlying idea was to combine native planting with campus beautification and at the same time introduce the idea of landscaping for resource efficiency. The native plants would require no special watering and would shade the administration building during the summer, reducing air-conditioning costs, but let sunlight in during the winter when their leaves dropped, thus reducing heating costs. We approached Jim Reeves with this proposal, and he again supported the group's initiative.

We organized a work party, planted the trees, and they thrived. Buoyed by this success, the following year we arranged with Reeves to plant a couple of young coast live oaks (*Quercus agrifolia*), one on the south side of the main classroom building and another just above the annually disced back slope of campus.

Within a couple of years, all of these trees were ten to fifteen feet high and quite healthy—but not out of the woods, so to speak. As it turned out, the campus director of environmental services (responsible for campus maintenance operations) at that time felt that the California sycamores were too sprawling and not sufficiently erect in their growth pattern. He further took the position that this problem could not be corrected by pruning or training. They had to be removed, he said.

The best compromise we could reach with him was an oral agreement to leave the most erect of the original plants, replace the other three with new California sycamores that satisfied his standards for good posture, and let us replant the "offending" specimens in a different location, where their sprawling postures would be less noticeable.

Despite this accord, the three least symmetrical trees were subsequently removed without notification and were never seen again. The "offensive" trees were replaced with nonnative London plane trees, a hybrid cross of two alien *Platanus* species from eastern Europe and the eastern woodlands of North America. The new trees are more erect in posture, but they are visibly a different kind of tree from the one surviving California sycamore. They also require more water and have comparatively less habitat value.

Off-Campus Activities

In part because of frustration with the outcome of campus projects, for a time we concentrated most of our efforts on off-campus projects. Thanks to the rugged topography and unstable geology of the Palos Verdes Peninsula, there are still a couple of thousand acres of undeveloped land in the area despite its proximity to Los Angeles. And as in many other parts of the country, our area has an array of organizations focused on environmental concerns, including native species and habitat. Most of these organizations are delighted to have volunteer help and to collaborate on educational programs to explain the ecological significance of their projects.

It has been fairly simple, then, to find off-campus projects offering opportunities for both community service and environmental education. My own volunteer involvement with a number of organizations has made it even easier to coordinate student participation and to give more committed students a chance to learn more about how such organizations work by serving as student representatives on the organizations' boards of directors.

In addition to these local projects, most students enjoy the opportunity to escape from the Los Angeles megalopolis from time to time. Whether as field projects for courses in geography or ecology or as service projects for GAP/MOVE, scores of students and at least a sprinkling of college faculty, administrators, and staff have volunteered to work on habitat restoration projects in national forests, Nature Conservancy preserves, state parks, and Audubon Nature Centers in California, Arizona, and Hawaii—the latter two in week-long "alternate spring break" projects.

Working with off-campus organizations is a treat for me because they are already knowledgeable, professional, and committed to habitat restoration. I have often been able to set up projects in such a way that we incorporate an orientation to the area by the site directors or managers, who also explain their personal backgrounds and some of the research activities and management tasks incorporated in their roles as preserve managers. For most students, exposure to these role models is an eye-opening experience, directly conveying the point that environmental issues are important, that there are lots of dedicated people hard at work on research, restoration, and management, and that there are constructive things that *they* can do to make a difference.

Campus Plantings, Round 2

Working on off-campus projects, however satisfying it may be, is not the most effective way to advance the environmental consciousness of Marymount College and the students who pass through it. I was therefore delighted in 2002, when several members of Marymount's student government—president Jun Miura, who had previously volunteered with me on a Nature Conservancy preserve, and freshman representative Lindsey Hill—were determined to celebrate Earth Week and sponsor several related activities, including an information fair and a speaker forum.

As one component of the week's activities, we tried another native planting project on campus, this time concentrating on wildflowers and shrubs. Once again Jim Reeves was supportive and helped identify an appropriate area for the project. The student government provided $220 for purchasing plants, and I contributed some plants I had propagated in my backyard. Although some did not survive—some plants are particularly sensitive to transplant shock when planted out of season, and the record drought in 2002 did not help—nonetheless it was at least a partial success, and we hope the surviving plants will provide a core around which new plantings can be established in the future. The overall effort was encouraging, and Hill commented, "I did not receive anything but support from administration, and was pleased at their eagerness for the positive changes I was trying to promote."

Student Reactions

Does participation in environmental projects have any discernible long-term influences on students? Subsequent assessments seem to depend on a combination of factors, including the tasks performed, working conditions, and the extent to which the rationale for the tasks was made clear. It is relatively easy to get students to participate in "life-affirming" activities like plant propagation (that is, unless they have fear and loathing of handling soil because it's "dirty"). It is often more of a challenge to recruit students to remove nonnative plants, because they may see it as life destroying rather than as enhancing prospects for threatened native species. Of those who volunteer unsolicited comments about the projects they participate in, the substantial majority are positive in terms of both the educational value of the experience and the satisfaction they feel in making a constructive difference. Many express a desire to participate in the same or similar activities in the future, although they do not always follow up when given the opportunity.

It is my impression—and it would certainly be consistent with research on attitudinal and behavioral change—that the long-term impact on students depends in large measure on their initial motivations for involvement in the projects. If their participation is purely voluntary, based on the intrinsic appeal of the course or activity, or a desire to do something constructive, or engage in outdoor activities, students generally enjoy the experience. If their participation is motivated primarily by a desire to gain extrinsic rewards, such a better grade in a class or credit for community service, their levels of enthusiasm during participation in the projects, and their subsequent assessments of the experience, are less predictable.

Institutional Impacts: Why Things Work (or Do Not)

I started at Marymount as a "green" professor in two very different ways. I was personally concerned about green ideas and practices, such as valuing nature and sustainability, and I was also naive, "green," as to the nuances of Marymount's institutional culture and the most effective ways to get things done. Advocating for sustainable practices in areas such as landscape management has been an education.

Some obvious lessons can be culled from the GAP/MOVE native planting projects. It is clearly important to understand competing priorities, such as the importance of greenscaping as a cosmetic marketing and community relations strategy in an institution like Marymount College. Also, technical constraints like county fire control regulations or city codes and permit conditions may dictate landscape practices, including such details as approved species and watering schedules (watering is restricted in some local areas because it may contribute to geological instability and landslides).

It is clearly helpful to develop relationships with others across campus, both to build grassroots awareness and support and to sustain dialogue with strategically- positioned people, as GAP/MOVE has done with the vice president of college operations Jim Reeves. Personal agendas and tastes can complicate matters, however, as in the case of the former director of environmental services, who found native sycamores incompatible with his landscape aesthetic.

Ongoing routines like maintaining native plantings or habitat areas require another level of planning and follow-through as compared to one-time events. Putting a plant in the ground is not the end of the task of growing plants, and all involved need to understand the need for long-term commitments. Someone has to assume responsibility for ensuring that information and appropriate actions are effectively communicated to others whose actions may have an impact on a project, as with the tractor operator who disced GAP's native plantings to reduce fire hazards.

The comparatively small size of Marymount College also seems to have a significant effect on efforts to advance environmental awareness and sustainable practices on campus. Everyone on campus is relatively accessible, and thanks to the school's small size, most college personnel know each other by name. In another sense, though, the school's small size can be a disadvantage. The degree of job specialization is not as great as at many other colleges and universities, and one consequence is that Marymount has no one specifically charged with formulating and implementing either educational programs or management practices addressing the issues of stewardship or sustainability. Although many people are sympathetic to environmental concerns and conceptually support good stewardship practices, few actively advocate for change, lest their efforts be construed as volunteering for additional workload and responsibilities. Everyone is already busy and has other priorities that are more central to their roles in the college.

The absence of any formal position specifically responsible for environmental policies and practices has contributed to a lack of consistency and follow-through for some of the GAP/MOVE initiatives that were intended to promote environmental awareness and sustainable practices.

Leadership in addressing this issue is coming from Marymount's student body. At the instigation of Lindsey Hill, the college's student government has recently addressed this accountability issue by creating a special position for an environmental representative in the student government. The first person elected to fill that position is Lindsey, who will set a high standard for her successors in the post.[2]

The day is coming when the college administration will follow the student government's lead and designate an environmental ombudsperson, and it will become routine practice to consider environmental sustainability in all college facilities and operations, including landscape and native habitat management. In the meantime, student leaders like Eddy Sison, Lupe Gonzalez, Eduardo Flores, and Lindsey Hill—who recognize that it is everybody's responsibility to be good stewards of air and water quality, soil, biological diversity, and other natural amenities—will help us get there.

Notes

1. Marymount College, Palos Verdes, is one of a family of semiautonomous institutions founded by a Roman Catholic order, the Religious of the Sacred Heart of Mary. There are five other Marymount colleges in the United States and others abroad. Whenever I refer to "Marymount College" in this chapter, I am referring to Marymount College, Palos Verdes.

2. During her first year in the new student government post, Lindsay focused her efforts on expanding recycling and then coordinated a meeting with the vice president of college operations, the director of environmental services, and others. She negotiated a promise of formal budget support for part-time student recycling workers, under the director of environmental services, and she has also lined up student government sponsorship of on-campus native plantings, in coordination with the college administration.

V

Building System-wide Commitment

13

South Carolina Sustainable Universities Initiative

Patricia L. Jerman, Bruce C. Coull, Alan W. Elzerman, and
Michael G. Schmidt

The University of South Carolina is in downtown Columbia, a block from the state capital. The university enrolls 15,300 undergraduate students and 9,000 graduate students. Clemson, the state's comprehensive land-grant university, with 14,000 undergraduates and 3,000 graduate students, is located close to the Appalachian Mountains, 140 miles northwest of Columbia. The Medical University of South Carolina, with approximately 2,300 students, is located in the historic port city of Charleston, 110 miles southeast of Columbia.

To those who are aware of the fierce football rivalries among South Carolina schools, the idea of intercollegiate cooperation seems unlikely at best. Thus, many are surprised to learn that sixteen of the state's colleges and universities have joined together in a loose coalition to promote sustainability in the classroom and in campus operations.

The South Carolina Sustainable Universities Initiative (SUI), as the group is called, is unusual in that there was no state government call for a focus on sustainability. Nor was there a crisis sparked by pollution, poor environmental management, or health and safety concerns. Instead, the cooperative venture began with the external stimulus of a private foundation with business interests in South Carolina. The European-based foundation had two goals: to serve as a catalyst for cooperation among the state's Research I universities and to use the universities as instruments of change to enhance understanding sustainability in the United States.

In South Carolina, each state-supported school has its own board of trustees, and all compete against each other for funds and students, thus limiting enthusiasm for working together. There is no overarching board of regents to enforce cooperation. Thus, it is doubtful that the SUI would exist without the impetus provided by the possibility of external funding.

Each university sent representatives to an initial meeting, where the idea of embarking on a joint project was discussed, but with little movement toward action. Several more meetings were held, and eventually the right mix of personalities came to the table, encouraged by development officers interested in securing funding. These individuals were personally concerned about sustainability or the environment, or both, and were willing to try to work with like-minded individuals from the other two schools. Of that group, three (Bruce Coull, Alan Elzerman and Michael Schmidt) became the leaders of the effort and hired a program manager (Patricia Jerman). We have developed a close and effective working relationship and are convinced that to a large degree, our friendship has facilitated our ability to manage the effort. Because we like each other, we talk more often and can handle disagreements and extreme honesty with grace. Each member of the team wants to support the others and make the entire initiative work as well as possible.

Bruce Coull, a marine ecologist, is the dean of the School of the Environment at the University of South Carolina (USC). The main campus is located in downtown Columbia, close to the state capitol, where it was established in 1801. There are eight regional campuses located throughout the state. USC has a medical school, law school, and a variety of graduate programs, as well as a large undergraduate enrollment.

Alan Elzerman is a Clemson University geochemist who chairs the Department of Environmental Engineering and Science, as well as Clemson's School of the Environment. Clemson is located in the hilly northwest corner of the state, on land donated by Thomas G. Clemson, John C. Calhoun's son-in-law, to establish an educational institution for the purpose of teaching "scientific agriculture and the mechanical arts." The school opened in 1893. Clemson is the state's land grant institution, with a large public service component, but it also is well known for programs in engineering and architecture.

Michael Schmidt is a microbiologist and a professor in the Department of Microbiology and Immunology at the Medical University of South Carolina (MUSC), located in downtown Charleston. Established in 1824, MUSC has grown from a small, private college training only physicians to a state university training a broad range of health professionals and biomedical scientists.

Patricia Jerman's background is in biology and public administration, and her experience encompasses a range of environmental work from serving as environmental advisor to former governor Dick Riley (who also served as secretary of education during the Clinton administration) to private consulting to directing an environmental nonprofit organization. These positions have resulted in a network of contacts, familiarity with how things are accomplished in the state, and an established track record in the environmental arena. All of these, plus a bit of chutzpah, are very helpful in functioning as a master's-level staff member in an environment where competence is generally assessed by the quantity of papers published and the size of grant awards.

First Steps

Fortunately, none of the schools in the SUI network has had a major environmental problem that focused attention on environmental improvement. The state has not issued a directive requiring government entities to consider sustainability, as some other states have. Although we lack the external incentives such directives would provide, we have had the luxury of defining our own goals and focusing on what we think will work best at our own institutions.

Without generous external funding, there would have been no motivation for the state's schools to work together. Within each school, the availability of funding worked to establish credibility; if a foundation cared enough about sustainability to contribute a significant amount toward achieving it, faculty and administrators were willing to take the time to hear what it meant. Once we had their ear, we usually had their interest.

We knew that we wanted to engage the entire university community—students, faculty, and staff—and that we wanted to improve our institutional footprints, ensuring that we were demonstrating best practices as well as talking about them. With generous financial support in place for a period of five years, we began by holding open meetings at each of the three universities to inform faculty of the initiative and solicit ideas about how to proceed.

What we found on each campus was fairly consistent. Sustainability, as opposed to environmentalism, was a hard concept to grasp; faculty were most interested in how sustainability could further their personal research agendas, and participants preferred to suggest changes that should be made by someone else. Nevertheless, in every meeting we had, there was at least one person for whom the ideas were new, and created an "Aha!" moment. They tended to be those whose field was not necessarily related to environmental issues but who had personal interests or a sense that change was needed in his or her own area of research.

As a result of these sessions, we decided that we needed to work hard to recruit faculty and administrators who were likely to consider sustainability with fresh eyes and were filled with the excitement and energy that comes with exploring new ideas. We tried to identify centers of energy and individuals who would be effective champions for our message. We also realized that with good ideas bubbling up in a wide variety of disciplines, we were not comfortable concentrating our efforts in one area, to the exclusion of others. Our favorite metaphor for our approach was to "light a lot of small fires and see which keep burning." We realized that key campus institutions and players were more likely to last multiple years than the SUI itself, and that our chances of institutionalizing the changes we hoped to make were far greater if we focused public attention on the relevant campus office or department rather than on the initiative. Finally, we recognized that not every strategy would work the same at each institution, making it important to know our audience. We came to understand, somewhat belatedly, that it is also important to know the strengths and weaknesses of our partners. We were unprepared for the degree to which individual personality and circumstance determined the success of our activities.

We settled on a program of mini-grant funding (in the $3,000 to $10,000 range) to engage faculty; a fund to encourage operational changes, with a required match to ensure departmental commitment; and student internships to carry out a variety of projects. We also planned a number of speakers, conferences, and workshops open to all of the state's institutions of higher education, as well as the general public. We were fortunate that our foundation support enabled us to fund these efforts, since the potential for funding attracts considerable attention. Nevertheless, the very existence of the SUI, with attendant opportunities for coordination and information sharing, has resulted in a number of positive changes that require relatively little funding.

In the rest of this chapter, we describe the expansion of the SUI, highlight some of the "small fires" that have grown into significant forces for change on campus, and we identify characteristics that make a successful champion.

University Collaboration

Although the SUI was initially focused on the three research universities, we wanted to include other schools in South Carolina whenever possible. To that end, representatives of all colleges and universities in the state were invited to our opening conference, held in January 1998. We invited faculty, administrators, and facilities managers and had a good mix of each.

Environmental issues have not been embraced in South Carolina to the degree seen in, for example, Vermont. Thus, we turned to a traditional opinion leader in the state—industry—to convey our message. A panel of well-known and highly respected industry representatives all delivered the same message: sustainability is important, and environment is important to the bottom line both intrinsically and as a source of new business ideas. From comments heard after the session, this outlook was new to many faculty members and helped them to view the sustainability initiative as a tool to make students more marketable.

To help our participants envision possibilities, we invited Kurt Teichert from Brown University's Brown Is Green program and Carol Carmichael from Georgia Tech to describe their programs. The advice we received from these experienced pioneers was invaluable. Carmichael cautioned us to be patient: change comes slowly, and we could not expect to achieve lasting change in a period of only five years. Teichert reinforced our sense that others should receive as much credit as possible by saying that as director of Brown Is Green, he tries to be "everywhere and nowhere."

A centerpiece of the conference was the signing of a formal agreement among the presidents of the three research universities. Although they had agreed to work together on the initiative, we felt it was important to have this agreement captured in a formal document. We considered the Talloires Declaration but decided to develop a state-specific document instead. Our reasons were both philosophical and practical. Because of the history of rivalry among the three founding universities, we wanted to focus on cooperation among schools in South Carolina and felt that a state-specific document would be a better vehicle to turn attention inward. On the practical side, we expected that it would be easier to persuade our presidents to sign a home-grown document that was less elaborate than the international Talloires Declaration. We were right. The document was signed at the opening conference of the Sustainable Universities Initiative, in January 1998.

Expansion to Other Schools

Representatives of other schools across the state had always been invited to our conferences and workshops, and many participated. However, there was no formal way to involve the other schools until 2000, when, with the help of our funders, we were able to secure a one-time $300,000 appropriation from the state's General Assembly. The presidents of all state-supported colleges, universities, and technical schools were invited to become SUI affiliates. Thirteen of twenty-three chose to do so.

The president of each school designated an SUI fellow, who received a small stipend to support sharing information across his or her campus. In addition, faculty and staff of the affiliates were eligible to apply for mini-grants to support new courses or environmental projects on campus. Participation varied widely across campuses, with some submitting many requests for mini-grants and others submitting only one or two.

The diversity of fellows appointed by each school allowed us to recognize characteristics of success that are undoubtedly relevant in most schools. The ideal campus coordinator is capable of leading opinions on campus, finding collaborators, and identifying young faculty members who could benefit most from mini-grant or other support. In some cases, the person named as fellow had a longstanding interest in environmental issues and was a logical choice. In others, the individual named was a young faculty member whose chairperson considered the SUI fellow appointment to be a career enhancement. Although these individuals certainly have energy and enthusiasm, they may not have the status or comfort level to change opinions or behaviors on campus, and as relative newcomers, they may not know other potential collaborators outside their department. It may be that the best candidates for leadership at any university or college are faculty who have achieved tenure and full professorship and are looking for new challenges.

In January 2002, we held another statewide conference for the purpose of celebrating success and sharing our activities with members of the legislature and other key individuals. Since two of the original three signatories to the Sustainable Universities Agreement had retired, we felt it was appropriate to hold a "recommitment" ceremony. To our surprise and delight, the presidents of the three founding schools were joined by eight other college and university presidents or their representatives at a press conference to mark the occasion (figure 13.1).

Are Coalitions Helpful?

We are often asked whether the advantages of working with a coalition outweigh the disadvantages, and that is a hard question to answer at times. As anyone familiar with large state universities knows, change is a slow and cumbersome process. When several schools are involved, the process becomes even more cumbersome. There are multiple approvals to seek, several bureaucracies to navigate, potentially conflicting procedures to follow. However, we have found advantages to working as a coalition that justify the effort.

One advantage is positive attention from lawmakers (who determine state school budgets) and other state officials who are accustomed to seeing the state's major universities compete rather than cooperate. Although the potential has not yet been realized, we believe we will eventually be able to use the coalition in informal collective bargaining or collective lobbying with state government. Officials from our schools believe that there are provisions in state procurement and engineering regulations that actively discourage, rather than encourage, environmentally prudent purchasing, construction, and management. While at least one of our member schools has been successful in changing some state regulations to accommodate green construction, we believe much more can be done and plan to develop systematically a list of regulations requiring review.

Finally, the most important advantage is the boost that information sharing gives to all schools. The SUI provides a mechanism to let one school know what another school is doing, assist faculty in finding allies either within their own school or at other schools across the state, and bring like-minded individuals together for workshops and meetings. Although it sounds obvious, we have been surprised by the number of positive outcomes that have arisen simply by bringing people together in one room. We sometimes joke that a budget sufficient to provide coffee and cookies or lunches may be one of the most important attributes of the SUI. We have also been struck by the number of times the SUI has been in a position to tell individuals at the same institution (occasionally

The South Carolina Sustainable Universities initiative represents an intellectual community committed to the advancement of theoretical and practical knowledge, as well as a collection of physical operations rivaling small towns in size and scope of impact on the environment. Recognizing our role as a positive force in South Carolina's economic and social advancement, we believe it is incumbent upon us to cooperate in leading the way toward a more sustainable future through our teaching, research, community service and facilities management.

We therefore singly and collectively commit to:

- Fostering in our students, faculty and staff an understanding of the relationships among the natural and man-made environment, economics, and society as a whole.

- Encouraging students, faculty and staff to accept individual and collective responsibility for the environment in which they live and work.

- Serving as a center of information exchange for other institutions within the state.

- Operating existing facilities and constructing new facilities so as to maximize efficiency and minimize waste, thereby protecting the environment and conserving resources.

Signed by the Presidents of:

Clemson University | **Medical University of SC** | **University of SC**

Central Carolina Technical College | Coastal Carolina University | College of Charleston

Francis Marion University | Lander University | Midlands Technical College

Trident Technical College | Winthrop University

Figure 13.1
Text of South Carolina Sustainable Universities Initiative

even within the same department) about relevant work or interests of a colleague. The ability to do this well requires what we think of as a spider web view of the project: the ability to see how various strands are (or can be) connected and to understand how a disturbance in one portion of the web can have an effect on distant strands. Individuals who are very focused on accomplishing a specific series of tasks will be less effective in this role than individuals who are focused on helping others define and accomplish tasks.

Another valuable aspect of information sharing is the boost that the ensuing subtle and gentle competition gives to the efforts of all schools. It has been interesting to watch mental wheels turn when officials at one school hear about efforts at another. Reactions have ranged from "can we come meet with you to find out how you did it?" to, "We're not going to let them get that done before we do." Either response serves to move one or more schools forward.

Examples of "Fires That Blazed" and Their Champions

Three principles have informed our approach to the SUI: find centers of energy, or "small fires," and help them grow, identify champions who have both the excitement and the standing to spread ideas across campus; and allow others to take credit for positive developments. Many small fires have taken off, ranging from the development of environmental management systems (EMSs) to a vermicomposting operation that has been visited by representatives from many states and Canada. Champions have emerged on all of the SUI campuses, and in many cases practices instituted by the SUI have become an integral part of university life. (For more details, see the SUI annual reports at <www.sc.edu/sustainableu>.)

The following examples illustrate just a few instances of the happy mix of effective champion with a small fire that grew, and in some cases, spread well beyond what we considered possible.

The Greening of Housing at USC

Gene Luna, director of student development and university housing and a professor at USC, once described himself as a "reluctant environmentalist." USC's president, John Palms, established an Environmental Advisory Committee (EAC) under the aegis of the SUI; Luna was appointed and attended a meeting more out of a sense of duty than personal interest. At the early meetings, he heard discussions of USC's water and energy use and the effect of both on budgets. Luna subsequently (and without prompting) told the EAC that the contract for washers and dryers was due to expire, and he was considering installing water- and energy-saving front loaders. He did, University Housing received positive publicity, and the university is saving approximately $20,000 per year. Luna says, "Involvement with our Environmental Advisory Committee was a catalyst to my growing interest and passion for issues of sustainability and environmentally friendly practices." Moreover, he has become one of the strongest proponents of the SUI message.

At the same time, he negotiated with SUI to share in the support of a newly created environmental manager for housing position. The manager, Michael Koman, instituted a number of additional developments, which continue to spread to other parts of campus. These include a very successful end-of-year collection of approximately 80 tons of reusable items from students moving out of the residence halls, establishment of student-run initiatives in the residence halls, and a green housing Web site.

The housing division also piloted the use of electric utility vehicles on campus, draping the first such vehicle in a banner proclaiming "zero emission vehicle" and unveiling it as the campus swarmed with parents helping their children move in at the beginning of fall semester. (Those in charge acknowledged that "zero emission" was not entirely accurate, but felt that a punchy tag line to attract maximum attention was an acceptable means to the end of getting campus citizens to think about emissions at all.) Now, USC housing has purchased several more electric vehicles, including a truck, and other divisions on campus are following suit.

Because Luna also presides over a number of divisions, including student health services and counseling, his ideas are spreading beyond housing. Recently, he requested an analysis of LCD versus conventional CRT computer screens, and after seeing the potential cost savings over time, instructed all divisions under his jurisdiction to move toward replacing conventional monitors with flat-screen LCDs. Because of the way energy costs are billed at USC, Luna's divisions will bear the cost of the new monitors but will not benefit directly from the savings. This seems to be a common problem and a widespread impediment to change that would ultimately decrease the institutions' environmental footprint.

Perhaps the most exciting aspect of Luna's conversion is his interest in green building. After Tom Battenhouse, director of housing facilities, attended an SUI conference on green building, Luna and Battenhouse decided that the next residence hall constructed should be green—green enough to warrant Leadership in Energy and Environmental Design (LEED) certification from the U.S. Green Building Council. Plans for the building are well underway, and it is likely that the building will be among the first LEED-certified residence halls in the world. What will make the accomplishment even more exciting is that the building will have been constructed using the budget allotted for a conventional residence hall, within the time frame set aside for a conventional building (because plans were well underway when the decision was made to go green). Luna notes that "what is really exciting is the opportunity to prove that sustainability is good business both short and long term, fiscally as well as socially."

The building has already produced several positive results, including a partnership with the local utility, which will provide state-of the-art monitoring equipment and assist with green power technology. Because the building will include classroom space, there has been an unprecedented merging of faculty and housing interests, with various groups of faculty working to take advantage of the building's unique features and location. Finally, the building has served as a spur to several other members of the coalition, who are rushing to develop their own green structures. In at least one case, the housing-utility partnership has served as a model for a similar partnership at another member school.

Classroom and Community Ripple Effect

Members of the SUI have combined classroom learning with service or experiential learning on campus or in the surrounding community. In several cases, the enthusiasm of the initial leaders has produced a cascade effect, spreading the impact far beyond what was initially contemplated.

Horticulture and English at Clemson One of the earliest projects funded by the SUI was proposed by Mary Haque, a professor of horticulture at Clemson University. She and several other faculty members engaged their combined classes (horticulture, landscape architecture, and sociology) in a project to design a sustainable landscape at a Habitat for Humanity community near the campus. Students designed and planted a landscape of native plants with high value for wildlife and low water and other maintenance needs. The property became the first (and perhaps only, to date) Habitat for Humanity home certified by the National Wildlife Federation as a Backyard Wildlife Habitat. At the same time, students were asked to consider their own lifestyles and to make at least one change, to be discussed in a reflective essay. Haque has included the "personal sustainability" element in many of her subsequent courses, including an introduction to university life course for incoming freshmen.

Because Haque is well liked, well respected, and continually engaging in "spider web" thinking, she has engaged the imagination of a number of other individuals, notably those in the Department of English. In 2002, she enlisted the cooperation of a professor who taught both freshman English and a technical writing course for more advanced students in a project to design outdoor classrooms for nearby elementary schools. English 101 students assisted in brainstorming and capturing ideas for the horticulture design students. They went a step beyond and explored sustainability on campus. Ultimately, the students became so energized they developed recommendations for change on campus and invited a representative of the university's facilities management team to join the class in a discussion. In the meantime, the technical writing students were developing grant templates to be used by the public school staff and parent groups to solicit funds for installation of the gardens once designs were accepted. Those of us involved with the SUI had no idea of the extent of the English classes' involvement until we heard the end-of-semester presentations by the horticulture students.

Word travels fast, and another faculty member responsible for a federally funded tutoring program, America Reads at Clemson, chose sustainability as the topic for shared experiences between Clemson students and elementary-age children from nearby schools. That project evolved to include hands-on work to grow and plant native plants to aid in restoring a damaged stream bank on the grounds of a new school. The efforts to date have resulted in awards, extramural funding, and positive attention in the media.

Most recently, Haque and English professor Summer Smith Taylor combined forces to identify campus clients with real projects related to sustainability with whom technical writing students could work. Clients included housing administrators interested in developing a green dorm, faculty planning a campus farmers market to showcase locally and organically grown food, and several facilities managers seeking assistance in publicizing recycling efforts. Students in the eleven classes developed over 190 separate deliverables, learned first hand about sustainability, and developed skills that will serve them well outside the classroom. Faculty and administrative clients found the project time consuming but felt they received something of real value in exchange. One faculty member said he would have expected to pay a private firm over $30,000 for the services provided free by students, and that the students probably did a better job since they were familiar with the campus.

The English professors hooked by Haque have become advocates in their own right and have also joined with English and humanities faculty at other schools across the state to share ideas and sources in a group loosely called "Green English."

English 101 at USC

One group of green English faculty is located at USC. Several years ago, the SUI enlisted the aid of geography professor Kirstin Dow, another "spider web thinker," to explore possible interest in sustainability among liberal arts faculty. One outcome of this exploration was the establishment of an environmental writing contest. Managed by teaching assistants for English 101, a course taken by approximately 90 percent of incoming students, the contest has grown over the years. In 2001, the managing teaching assistant, Corinna McLeod, and faculty member Christy Friend expanded the project to designate nine sections of English 101 as environmental sections with a community service component. Six new teaching assistants (TAs) were exposed to environment and sustainability as a theme for writing courses, and approximately 220 first-year students provided 2,200 hours of service to thirty community or campus agencies. The program was doubled in 2002. We think the most important result of this effort is that each of the TAs will incorporate the approach they learned at USC into the classes they teach when they complete the Ph.D. program and embark on their careers. The department realized the value of the approach when one of the graduating TAs reported that the first questions he was asked in an interview had to do with his environmental service-learning work. The project has been featured in an article in the newsletter of the National Resource Center for the First Year Experience, which generated a request for more information from at least one other large state university. We hope the positive outcome for students and the national attention will ensure that the environmental sections of English 101 become institutionalized within the department.

Reforming the Medical School Curriculum At MUSC, the larger of the state's two medical schools, a member of the SUI steering committee has been instrumental in producing cascading beneficial results within the pediatrics curriculum. The director of pediatrics, J. Routt Reigart, has long been an advocate of children's environmental health, serving on numerous national advisory committees focused on children's exposure to lead and other environmental hazards. Reigart identified a promising new pediatrics faculty member, James R. Roberts, and suggested that SUI offer him partial support in order to allow him the flexibility to integrate environmental issues into the pediatrics curriculum. Roberts was convinced that pediatricians do not receive enough training in environmental threats to health to allow them to take a useful patient history. As Roberts said in a recent interview, "Environmental health risks are a leading cause of illness, due, in part, to the widespread use of pesticides. . . . Unfortunately, the average health professional receives minimal training in environmental health." He began remedying this in his own classes. At the same time, he has become a national expert in his own right, authoring (with Reigart) the fifth edition of EPA's *Recognition and Management of Pesticide Poisonings*, available in both Spanish and English. Roberts was featured in an hour-long Discovery Channel program on the dangers of pesticides for children. Together, Roberts and Reigart are training the next generation of clinicians to understand environmental health by nurturing a group of pediatric fellows who will be trained to understand environmental health issues and to include them in their research portfolios, and who will subsequently include them when they become faculty members at other institutions.

Common Threads

Although we wish we could identify quantifiable factors and replicable procedures to ensure success, for the most part, we cannot, but we have seen some consistent threads. We have found it very beneficial to be open to opportunities rather than to specify well in advance precisely what we intend to do. This has allowed us to build on existing successes and take advantage of what we have learned. It has also allowed us to capitalize on new information and opportunities. For example, when we initially envisioned likely points of impact, we planned to offer educational programs within residence halls. The few that were offered were flops. Yet we had completely ignored the potential for operational changes within the housing division, and these have proved to be very effective in reducing the university's footprint and informally educating students.

There are some hazards inherent in defining a program as it develops. When funding depends on grants, it is essential that a clear program of action be developed so that funders know what they are being asked to fund and evaluators can determine whether goals have been met. Although we have strayed from our original plans in many cases, we believe the actual outcomes have been better than we anticipated. We have been fortunate in having funders and evaluators willing to look at actual outcomes rather than intended activities.

Another lesson is that we need to know our audience (be it one professor or an entire facilities division) and tailor our approach to individual circumstances. As Congressman Tip O'Neil is reported to have said, "All politics is local." Similarly, we have found that to make an impression, we need to make sustainability local in the sense that we need to explain to our audience how our message is relevant to their work. This has been especially important at MUSC, where students hoping to become physicians or other health professionals are intensely focused on meeting very tightly prescribed courses of instruction. In the larger sense, we need to understand how to tailor our message to a South Carolina audience—hence, our reliance on links with business and industry.

Finally, in most, if not all cases of results exceeding expectations, credit lies with the personal attributes of the champion. This applies to the SUI leadership as well as to individuals on each campus. In the examples used here, Luna and Reigart occupy positions within the university hierarchy that ensure their viewpoints will spread. However, authority is the least of the attributes necessary for success. In the instances we have noted, the proponents of change have the personality and style that make them ideal candidates to evangelize. They are both liked and respected on campus, and they understand how to work within the system to persuade others to their point of view. They also understand the value of public relations and have used publicity to bring about even more positive results. Finally, they have an innate ability to see potential relationships and linkages that allows them to extend their reach in creative and interesting ways.

At the same time, it is important to note that some of our major disappointments have come about as a result of the personality or style of the identified champion. In some of these cases, the individual was full of enthusiasm but unwilling or unable to engage others; the result was that the project did not take place or did not grow beyond its initial outlines. In a few cases, a champion stepped forward who had what might be called a "toxic personality" that prevented an excellent idea from coming to fruition. In most cases, the individual was the logical person to carry out a particular project, so it was difficult, if not impossible, to avoid working with him or her.

Identifying champions who have both the right personality and an appropriate position within the university community has been, paradoxically, one of our greatest challenges and one of our greatest pleasures. It is the single factor most responsible for success or failure and for making the difference between a good idea that becomes institutionalized and one that is a flashy one-time event.

14

Challenges of Greening a Decentralized Campus: Making the Connection to Health

Polly Walker and Robert S. Lawrence

Johns Hopkins University is a privately endowed, research-focused university located in Baltimore, Maryland. The university has a decentralized administrative structure for nine divisions in five locations. The Arts and Sciences and Engineering Schools are located on the 140-acre Homewood Campus in a residential area of row houses, apartments, and single-family homes. The East Baltimore campus, home to the Schools of Medicine, Public Health, and Nursing, is about four miles away in an impoverished area of row houses. In 1876, the hospital was located in this section of the city by its founder, Johns Hopkins, in order to serve the poor.

We are living in an aquarium called Earth. Ultimately, our common air, water, and climate link us all—our health and our very lives. Universities have a crucial role in educating students about these links, the importance of individual responsibility and the value of collective action, and engaging the students' imagination and creativity to find ways to correct the current negative balance between humans and nature. Greening the university means including the current and future environmental consequences of our actions in all aspects of university life. This is a complex challenge for large universities such as Johns Hopkins University (JHU), with many campuses and a decentralized administrative structure. Each of the campuses and each of the schools presents its own challenges and strengths for environmental stewardship. Decentralization makes sharing information about program successes and working toward one goal challenging.

Johns Hopkins has a special responsibility for greening its operations and curriculum because it is a premier health institution with a world-renowned hospital and schools of public health, nursing, and medicine. The vital connection between health and the environment is important for both greening a university and for informed decisions by health professionals. This connection is already clear for those committed to the ideals of sustainable living. At the same time, many health professionals working on specific disease entities may not be aware of either the enormous health implications of environmental degradation or how important their voices are in the debate. The imperative for action now is increased by restating the goals for greening the campus—not only in terms of the environment but also in terms of human health. Greening the university provides an opportunity to increase dialogue on these topics, especially at a university where health professionals work, study, and care for patients or develop population-based interventions.

As the largest private employer in the state, Johns Hopkins University also has an opportunity to set an example and influence policy by improving standards of university practice in recycling, energy use, food services, purchasing, green building practices, and transportation that minimizes environmental impact. Greening is not a trivial pursuit to be relegated to volunteers and do-gooders. Rather, universities are excellent places to begin establishing goals for institutional action and teaching based on the principles of environmental stewardship and sustainability.

Institutional philosophy changes very slowly and is often precipitated by individuals who are advocating for action. A series of steps, each propelled by individual students, staff, and faculty members, led ultimately to the creation of the Initiative for the Greening of Johns Hopkins University. As physicians who work in public health, both of us, Polly Walker and Bob Lawrence, viewed environmental stewardship as a health issue before coming to Johns Hopkins. Building on existing programs, we helped launch a more coordinated greening effort at the university. This chapter recounts the efforts that led to the initiative and explores some of our challenges in bringing a more visible and coordinated commitment to environmental responsibility in a decentralized university.

Getting Started

While some staff and faculty members at several campuses throughout Johns Hopkins were working on various environmental issues, there was no university-wide or even divisional effort to decrease the university's environmental impact and, more important, no university policies to guide such actions. Since actions were not coordinated or communicated to the university community, many of the projects or accomplishments were unrecognized. For instance, cost savings analyses and Environmental Protection Agency requirements prompted energy-efficiency programs including window replacement, energy-efficient lighting, and new energy-efficient buildings, but very few people were aware of these efforts.

Students provided the impetus for the first increased attention to green practices at the university. In the mid-1990s, undergraduate students protested the lack of an on-campus recycling program. University leadership responded by creating the position of recycling coordinator in the Facilities Department, and Pat Moran was hired in 1996. Pat's dedication to recycling and environmental stewardship has been crucial to what has been accomplished in the greening movement throughout Johns Hopkins. He enthusiastically set about creating a successful recycling program on the undergraduate campus by first tracking the proportion of trash that was recycled and then modifying procedures to increase the amount. Recycling programs did not spread to other campuses, however, because the recycling coordinator was isolated in one location and there was no university-wide goal or policy for recycling.

Bob Lawrence joined the Johns Hopkins faculty as the associate dean for professional education at the School of Public Health in 1995. His training and practice in clinical preventive medicine and his commitment to human rights and equity led to his appreciation of the environment as the ultimate public health problem. When a funding opportunity arose in 1996, and with the encouragement of the dean of the School of Public Health, Bob created the Johns Hopkins Center for a Livable Future (CLF) to advance thinking and action on issues of environmental sustainability, health, and equity, with particular emphasis on the links among health, diet, food production, and environment.

Polly Walker joined CLF as coordinator in 1996 after many years as an advocate for local and state environmental issues, including land preservation, drinking water quality, recycling, and air pollution. As she began work at CLF on issues of sustainability and health worldwide, it soon became apparent that there was only minimal attention to environmental stewardship at the university itself. This disconnect was highlighted by questions directed to CLF by students, staff, and faculty: Why isn't there more recycling? Why aren't incentives in place for alternative transportation? Why is so much polystyrene plastic used in the cafeterias? In trying to answer these and other questions, Polly began to look for ways to apply CLF's goals of sustainability and stewardship in a practical and local way at the university.

CLF began its programs by sponsoring interdisciplinary research and symposia on the health and environmental consequences of methods of food production, energy use, waste disposal, and consumption. Its actions stem from a recognition that decreasing the total consumption of energy and goods in the developed world is as important to the preservation of the world's ecosystems as population control is in the developing world. For instance, the average baby born in the United States today will consume over its lifetime more energy and goods than twenty-three babies born in India.

Having an organization on campus already dedicated to issues of sustainability and living lightly on the earth was a great advantage in initiating university greening activities. CLF's advisory board, comprising faculty members from four schools (Public Health, Engineering, Arts and Sciences, and Medicine), facilitates university-wide discussions on sustainability. In addition, CLF staff members have provided organizational and administrative assistance to the greening committee during its first few years.

In 1998, Pat Moran, the recycling coordinator, scheduled a meeting with Polly Walker to find out about CLF's activities. We discovered that both of us had been working toward the similar goal of reducing the environmental impact of the university. We both believed that the university was an ideal place to teach environmental stewardship to the next generation and to set an example of ecologically sound practices. Joining forces to work together toward common goals was an important turning point for greening at Johns Hopkins.

Educating Ourselves

Meeting like-minded advocates from other universities was the next important step in establishing a more concrete greening effort at Johns Hopkins. At the instigation of a graduate student, Karen Stupski, a group from Johns Hopkins attended a November 1999 workshop in Connecticut sponsored by Second Nature, "Education for Sustainability—Shaping the Future: Best Practices in Higher Education." Funding for the five conference participants (including Pat Moran, Karen Stupski, and Polly Walker) was provided by the Dean's Office, Facilities Department, and CLF. The conference provided valuable, practical information about greening experiences of other campuses and facilitated contacts at other academic institutions. Since the conference, the experts we met there have given us much valuable advice, and several were later invited to speak at Johns Hopkins.

After the conference, the Johns Hopkins participants met with Bob Lawrence to discuss new ideas for creating a greening program. Bob suggested convening a voluntary group of students, staff, and faculty to discuss how to increase greening activities at Johns Hopkins and to strengthen the commitment of the university leaders to this goal. We invited faculty, students, and staff with an interest in environmental stewardship and all those who had previously sent questions to the CLF about environmental problems at the university. This volunteer greening committee of about twenty faculty members, staff, and students met regularly to discuss facilities issues such as energy, transportation, purchasing, recycling, and transportation. We also discussed greening the curriculum by adding courses on sustainability and including the concept of living sustainably as a core part of education at the university.

The greening committee began its work by educating itself on the current status of greening activities at Johns Hopkins, researching programs at other academic institutions, and mapping out an appropriate strategy. In addition to research and following up on contacts from the Second Nature conference, we also participated in the National Wildlife Federation's Ecodemia program that provides materials and consultation to groups on college campuses attempting to increase environmental practices. The Ecodemia program consultant offered expertise and insights from the experiences of other colleges and universities that had already instituted or were attempting similar greening activities. We decided that the most important first steps were to obtain a clear signal of support from the president of the university and to convene a conference at Johns Hopkins.

The committee approached William R. Brody, president of Johns Hopkins University, and presented the issues we had been working on. He responded by sending an e-mail message to the entire university commu-

President Brody's E-mail, Earth Day April 21, 2000

Subject: Earth Day
Date: Fri, 21 Apr 2000 16:37:49-0400
From: "William R. Brody"

April 21, 2000

Dear Colleague,

Saturday is the 30th anniversary of Earth Day. For many Americans, the first Earth Day in 1970 was the first occasion that really underscored the magnitude of the environmental problems facing humanity. At the start of the new millennium, it is even more apparent that how we use the earth and its resources will determine the kind of earth we leave our children and our children's children. We face an enormous challenge: to protect the natural world and, at the same time, meet the needs of the world's growing population.

Universities can help society meet these challenges by forging new knowledge and providing students with the necessary tools to solve problems. Most Americans consider themselves environmentally minded, and many are quite conscientious about acting, as best they know how, on their environmental convictions. But few of us fully commit to conduct our lives in ways that preserve resources and enhance the environment. I believe that behavioral and cultural changes in individuals and society are necessary in order to create sustainable solutions for our shared future. Universities must lead the way in discovering how to bring about such changes.

The university can also set an example of positive behavior change and commitment to the environment and sustainable practices. I, therefore, announce the "Greening of Johns Hopkins University" initiative. Through this initiative, we will bring an environmental ethic to the university's operations, including procurement and services. The aim will be to help create a sustainable future. We have already begun this process - replacing light bulbs with more energy-efficient lighting, for example, and recycling paper, establishing a recycling coordinator position at the Homewood campus and initiating an inventory of current practices at Homewood and the School of Public Health. This is just the beginning. With the leadership of committed faculty, students and staff, Johns Hopkins will now work on substantially reducing the university's footprint on the environment.

I hope all of us will take some time at this observance of Earth Day's 30th anniversary to reflect on these issues. You will be hearing more about the "Greening of Johns Hopkins" soon.

Sincerely,

Bill Brody

nity to celebrate the thirtieth anniversary of Earth Day, April 2000 (see the box). The president's message provided critical support for decreasing the university's ecological footprint and including sustainability in its business practices and curriculum development.

The Ad Hoc Committee for the Greening of Johns Hopkins University

As a result of the president's message and with his support, the Ad Hoc Committee for the Greening of Johns Hopkins University was created, with Bob Lawrence as the interim chair. The members of this committee included the volunteer group of faculty, students, and staff who had already been meeting on greening issues, as well as others who learned about the initiative. This volunteer, unofficial committee has been meeting for three years on a regular basis, sharing information and seeking ways to effect change.

In order to focus our efforts, the committee decided to concentrate on just two campuses: the undergraduate campus at Homewood and the School of Public Health. These two campuses were selected because important steps had already been taken there. We formed subcommittees to work on greening issues at each campus. Recently, the Ad Hoc Committee has been greatly strengthened by the active participation of the new university facilities manager, Larry Kilduff, who is committed to ideas of sustainability.

An October 2000 conference, "The Greening of Johns Hopkins: Present and Future," was the committee's first campus-wide activity. This one-day conference was organized by the Ad Hoc Committee in order to increase awareness about greening and to select priority issues. President Brody gave the welcoming address, and David W. Orr, chair of the Environmental Studies Program at Oberlin College, was the keynote speaker. Reports from university staff on activities in recycling, purchasing, energy, and transportation set benchmarks for measuring future success. Faculty and student perspectives were also presented. The 100 attendees from many parts of the university then participated in breakout sessions on specific topics to determine future goals for greening the university. Each group brainstormed the possible greening activities and then chose its top two priorities for the university.

One suggestion from the conference was to expand the baseline data on current environmental practices. Surveys were conducted at both the Homewood campus and the School of Public Health to obtain benchmark information about current environmental practices, such as recycling, purchasing, and transportation. The Homewood survey was sent to departmental administrators, and the Public Health survey was sent on-line to all staff, students, and faculty. In addition to assessments of current efforts to address priority areas, the surveys elicited many useful suggestions for future action and helped publicize sustainability issues to the broader university community.

Since most people on campus were unaware of the university's actions to decrease energy use or increase recycling, another important early step was to publicize past and current efforts by the university to decrease its footprint on the earth. A Web site was created for recycling and greening activities at <http://www.jhu.edu/~recycle/greening.html>. A listserve was created to encourage personal action by forwarding articles of interest, research opportunities, and events about greening and environmental stewardship. An additional e-mail list links those interested in biking to work or school. Suggestions for personal actions are written by a member of the Ad Hoc Committee and published in the "Green Tips" column that appears periodically in *Human Resources Today,* a bulletin for all JHU staff, and the *Johns Hopkins Gazette,* a university-wide weekly newspaper. A logo, created for the October conference, continues to draw attention to the committee's work. Since the JHU mascot is the Blue Jay, committee member Royce Faddis designed our logo as a Green Jay (see fig. 14.1).

In addition to sharing information at meetings and via the listserve, the committee seeks other means of becoming educated about green practices at other institutions. For instance, members of the Ad Hoc Committee and several facilities administrators toured the Chesapeake Bay Foundation's new Philip Merrill Center, the only office building in the world to be platinum certified (The highest of four certification categories, requiring a score of 52 or more out of a total of 69 points) by the U.S. Green Building Council's program, Leadership in Energy and Environmental Design (LEED). The Chesapeake Bay Foundation is a nonprofit group dedicated to improving the health of the bay and its watershed. This tour provided us with much valuable information and

Figure 14.1
Logo for "The Greening of Johns Hopkins" conference

new insights on green building design and construction of particular interest to our facilities staff.

The School of Public Health's Official Greening Committee

While the Ad Hoc Committee on the Greening of the University continues as an unofficial, though university-sanctioned, volunteer group working on university-wide issues, the School of Public Health now has its own official greening committee. In spring 2002, Al Sommer, dean of the School of Public Health, appointed the Environmental Stewardship Committee, an official standing committee of the school, composed of representatives of all academic and supporting departments. The commitment and leadership of the School of Public Health are both logical and important, since ecological and human health are so closely linked.

The school's committee will begin by working to improve recycling and to increase green purchasing. Messages about environmental stewardship were sent by e-mail to all incoming new students in 2002 and 2003. In order to increase awareness, the committee and facilities staff developed brochures that outlined current programs and achievements and provided information on individual actions for recycling, green purchasing, and alternative transportation. These were distributed to all incoming students and are available to current staff and faculty members.

Challenges for the Future

The most important challenges for the Ad Hoc Committee on the Greening of Johns Hopkins are to sustain momentum, extend its work to the other divisions, in particular the medical school and hospital, and institute curriculum change.

Sustaining momentum is always difficult and requires long-term commitment by individuals to organize, plan meetings, research and publicize issues, and educate everyone on appropriate actions to take. Time constraints are a constant barrier to increased success. The recycling coordinator at the undergraduate campus is the only employee whose full-time job is greening the university. Since the Ad Hoc Greening Committee is a volunteer activity, greening activities are done in addition to normal job requirements or on personal time. Students are very interested but have many other demands on their time, and some degree candidates at the School of Public Health are in Baltimore for only one year. Larry Kilduff was hired as the director of facilities for the university in 2000 and has been cochairing the Ad Hoc Committee for the past two years. His interest in environmental stewardship and his responsibilities for new building and current facilities at the university make his role crucial for greening operations. Additional support by central administration would help sustain the committee's interest and commitment.

Extending the greening activities to other campuses and divisions is an important but very difficult next step. The medical school and hospital are particularly complex and challenging at Johns Hopkins and at other large universities as well. The primary mission of the hospital is healing the sick. However, the medical principle, "First, do no harm," is an imperative to consider practices that may in fact be contributing to disease directly or indirectly. For instance, mercury from discarded medical equipment accumulates in the environment and causes neurological damage. There are now programs that help hospitals become mercury free by either removing all sources of mercury or by careful accounting and recycling of all mercury used. Johns Hopkins has instituted some of these steps through its Environmental Safety Department. Another important step is to minimize incineration of hospital waste containing polyvinyl chloride because it produces dioxin and dioxin-like compounds that pollute the air and water and contribute to an increased risk of cancer and other disorders.

Another challenge is to institute curriculum change to include issues of sustainability for all undergraduates and public health students. The goal would be for students to have a clear understanding of the connections between the environment and human health and also about limits—limits to resources, limits to earth's capacity to renew, and the need for limits on what we discard and pollute. Greening activities such as recycling and energy efficiency are still considered interesting and even useful activities, but are not yet connected intellectually to the educational experience. Rather than relying on remediation and engineering to resolve harmful situations once they have been created, the emphasis should be on prevention and on innovative solutions. As important as these topics are, the numerous other requirements and demands of particular degree programs and professional education present major obstacles to greening the curriculum.

The most crucial challenge is to gain increased university commitment to the principles of environmental stewardship and to reducing the university's ecological footprint. A permanent university-wide committee appointed by the president of the university would be a major step in increasing the visibility and the effectiveness of the Ad Hoc Greening Committee. Most likely, many of the current Ad Hoc Committee members also would serve on the new committee. We believe that formal university appointment and a reporting responsibility to one of the top administrative officials would empower the committee to accomplish much more. The Diversity Leadership Council, which deals with issues of inclusiveness at all divisions, is a model for the type of university-wide committee we are recommending.

Part of the charge for the new official committee would be to develop incentives that could be instituted in order to decrease dependence on single-occupancy vehicle transportation, reduce energy and water use, increase recycling rates, and increase purchasing of green products and sustainably produced food. Many of these practices are extremely difficult to change one department or one division at a time. In order for real progress to be made, the university needs to incorporate green policies in all aspects of its mission.

Conclusion

Working to change large institutions is a slow process, especially in a decentralized system. There has been progress at Johns Hopkins. What is needed next is a commitment of funds from the university and an official university-wide committee dedicated to decreasing the ecological footprint of the university. It is especially important for this process to build on the university's strengths and resources. Linking sustainability and ecological responsibility to human health is an essential concept for the preservation of humankind and can serve as an important stimulus for progress. Johns Hopkins, with its status as a premier institution in medical and public health research, education, and practice, has both a unique potential and a unique responsibility to make decisions for facilities, purchasing, transportation, and curriculum that reflect stewardship of the environment.

15

Sustaining Sustainability: Lessons from Ramapo College

Michael R. Edelstein

Ramapo College of New Jersey, a state-supported four-year college of liberal arts and professional studies with a student enrollment of nearly 6,000, opened in 1971 with a unique interdisciplinary mission. It has since added to this mission the goals of offering a global, multicultural education employing an experiential approach and a focus on sustainability. The 300-acre campus is located in northern New Jersey at the gateway to the scenic New Jersey–New York Highlands region.

This chapter details some of the lessons from a thirty-year experiment in finding sustainability at Ramapo College of New Jersey, illustrating the richness of the academic experimental playing field. At their best, colleges and universities are the logical loci for experiments in sustainability. They are inherently learning centers, where new ideas can be heard and different interacting actors can advocate for change. They guide other sectors and have the potential to serve as societal models. Although they are unique in many ways, their innovations are largely replicable by other institutions. And their mission and responsibilities for defining educational scope give them a reason to update and reevaluate in a way that invites institutional learning and openness to sustainable thinking. Furthermore, colleges and universities are multifaceted microcosms, offering the potential to address sustainability through multiple convergent axes of change in four interconnected sectors—what I call "the four C's": curriculum, culture, campus, and community. There are comparatively more and more diverse footholds for making change in academic settings than in most others. Constraints also exist in academia, including the complex organizational, institutional, social, and cultural defenses against change that all institutions deploy. Meta-change of the type represented by a shift of society to sustainability is particularly challenging because of the general lack of critical social evaluation and learning mechanisms for understanding complexity and adjusting to the big picture, a limitation that bureaucratic and disciplinary centers of education are prone to.

Ramapo College of New Jersey was created in the late 1960s as part of New Jersey's effort to keep students in the state by upgrading its six teachers' colleges and creating two new interdisciplinary colleges, Stockton in the far south and Ramapo in the north. The interdisciplinary zeitgeist of this period was particularly expressed in the United States by the National Environmental Policy Act (NEPA) of 1969, signed on New Year's Day 1970. Seeking the preservation of a balance between human activities and the environment, NEPA recognized that this relationship requires a new mode of interdisciplinary thought and practice, uniting the arts, sciences, and social sciences. From the outset, Ramapo College was attuned to the environment by virtue of interdisciplinary milieu, organization, and mission, as well as its location facing the verdant Ramapo River valley below and the majestic Ramapo Mountains beyond, at the gateway to the highlands region. This unity was expressed in the campus design, built around natural and historic features of the site—an old estate and cattle farm—and utilizing mirrored windows to minimize the visual presence of the buildings while reflecting the mature trees and natural landscape.

As a new interdisciplinary institution with a unique and urgent mission, there was a comparative openness to experimentation among the young and often unconventional faculty attracted to the college. As one of these pioneers, now grayed over thirty years, I have enjoyed a stimulating educational environment. Although it has often competed with my research and writing about the human costs of environmental contamination, I have played a key role in one of the most important of these experiments: the effort to make Ramapo a sustainable institution. In this chapter, I recount my own take on this continuing and evolving project and reflect on some of the lessons. I observe that sustainable change is a fox-trot, often advancing incrementally through a combination of big and mincing steps forward in an effort to recover from periodic moments of retreat. Although reshaping long-hallowed halls must be even more daunting, in converting even young institutions to a sustainable path, one must be prepared for a one-step-forward, two-step-back rhythm.

Early Vestiges of Sustainability

Ramapo's evolving curriculum reflected its founding milieu. Three different environmentally oriented majors were offered, and an early reorganization in 1974 resulted in the creation of the School of Environmental Studies. The college's flagship environmental studies major soon attracted hundreds of students. Specializing in issues of a science-integrated social ecology, the major was comprehensive and sustainability oriented long before the term became a mantra. When the energy crisis hit in 1973–1974, faculty and students petitioned the first president, George Potter, for a parcel of land on which to build an alternative energy center. The center was inspired by the vision of social ecologist Murray Bookchin. It was given form and substance from the onset by Bill Makofske, a former nuclear physicist converted to the study of renewable energy. With continuing involvement of students, faculty, and alumni, the facility grew from an initial greenhouse building to an off-grid complex of student, faculty, and alumni-built structures that demonstrated all the elements of an ecological settlement except housing. Students built and then studied in the solar school house; grew seedlings and winter crops in the passive solar greenhouse; produced compost and raised enough vegetables and fruit in the French intensive organic gardens to supply area soup kitchens; practiced permaculture and sustainable tree planting on the grounds; raised chickens for their eggs, insect control, and manure; honed their skills in wind power on the several wind generators; and pioneered techniques for source-separated recycling in a state-of-the-art recycling shed that, before New Jersey's comprehensive recycling law, served the larger community and earned money to support the center.

Beyond its use for teaching and research about appropriate technology, the Alternative Energy Center (AEC) served also as a community center, hosting several annual community festivals for the college, including Sun Day, Earth Day, and a harvest festival. Events included apple bobbing, cow pie tossing, rope pulling, many scores of potluck dinners, and presentations by countless speakers, as well as workshops and conferences. Environmental and sustainability education was offered for regional schools, citizens of surrounding communities, and Ramapo students.

As a key player in making the dynamism of the Alternative Energy Center radiate throughout our program and institution, I felt part of an egalitarian community that included not only my colleagues but also my students. We were positioned to do the real work of blazing new trails away from an oil-based, overconsuming, and unsustainable way of life. There was no challenge that we would not confront. Collectively, we probably offered more innovative courses on renewable energy, social ecology, and the like than any other undergraduate institution of the era. The peer tutorial program I created and directed employed our best juniors and seniors in acculturating new majors within our active learning community. We also had the flexibility to pursue inspiration and creativity. For example, as the result of a discussion with students about what ecological housing would be like, I collaborated with two other faculty to teach a full-time, full-year academic program resulting in the design of an ecological living facility (ELF) for the Alternative Energy Center. Our work on the ELF was so infectious that Ramapo's founding president, George Potter, became a renewable energy advocate. However, instead of building a twelve-person solar residence, in the late 1970s, he determined to make the college's first high-rise dormitory solar heated. With this increasing institutional support, we were on top of the world.

This indication of our success soon proved an omen of challenges to come. The bold attempt to build a solar-heated five-story dormitory proved a flop due to design errors. When the building was ignominiously converted to gas heat, it left behind scars that discouraged experimentation with green architecture for a long time. Then the Reagan era in American politics occasioned a massive student stampede to the new School of Business, sucking environmental studies so dry that the school was closed and the faculty dispersed to other schools. Funding for the Alternative Energy Center also disappeared. Although the college continued its environmental majors and the Institute of Environmental Studies was formed to connect the dispersed faculty, both Ramapo and the larger society now appeared to view the environment—and, by extension, sustainability—as a fad that was over rather than as a new paradigm that had just begun. Finally, just as he had become our solid supporter, President Potter was forced into retirement.

Broadening Sustainability throughout the College

By the mid-1980s, Ramapo's new president, Robert A. Scott, had decisively moved to concentrate the college's energies on global education. The environmental faculty, our numbers diminished from fifteen to eight by nontenuring and attrition, now faced competing task demands and opportunities from our new institutional homes, further eroding the available energy for the program. But we were a resilient and dedicated group, and we determined to see what we could learn from our fall from grace. Recognizing that rebuilding the major to its former size and regaining school status were complex organizational tasks with serious roadblocks, we chose an alternative path to make environmental and sustainability issues integral to the college's broader educational enterprise. We secured our ability to support a smaller environmental studies major by gaining approval to list many of our core environmental courses as meeting requirements for a new all-college general education program. This approach not only guaranteed enrollment in our fundamental classes but also created a mechanism for recruiting majors. The approach offered another benefit as well: diffusing environment and sustainability into the broader college curriculum. Diffusion was to serve as the new guiding concept for greening efforts at Ramapo.

In October 1990, the Talloires Declaration of University Presidents for a Sustainable Future was issued. The declaration called on universities to use every opportunity "to address the urgent need to move toward an environmentally sustainable future," and it set forth ten guiding principles (Tufts University 1990). When I sent President Scott the newly issued declaration, he recognized the fit with his global education focus and signed enthusiastically, marking Ramapo as an early signatory. Having hooked President Scott, I also became hooked: his commitment became my responsibility. Seeking to move Ramapo toward our newly assumed goals, I sought ways to further the strategy of institutional infusion. As a result, I refocused outside the sphere of the environmental programs and moved into a whole-institutional role.

External Funding Builds Momentum

Learning about the educational innovations program of the U.S. Department of Education while on a subsequent sabbatical, I developed a proposal to the Fund for Improvement of Post-Secondary Education (FIPSE) to infuse Ramapo's broad curriculum with environmental literacy, which I defined as the knowledge and wisdom to create a sustainable world. The project was funded and ran from 1994 to 1998. Before pursuing this project, I had sought strong faculty as well as administrative backing. In fact, it won the unanimous support of the Faculty Assembly. Over the project's four years, more than half the faculty willingly participated, working through their curricular groups under supervision of my environmental studies colleagues to infuse materials relating to the environment and sustainability into courses and programs. In all, more than eighty faculty participated in approximately sixty-seven curricular diffusion projects, exceeding initial targets. Participants were broadly distributed across the curriculum and received a small stipend. The Alternative Energy Center now became an all-college resource in a new way, hosting sustainable education programs for all first-year students: forty-four tour programs were held for college seminars.

Most important, the effort was seen as a collective success, reflected in the response to a video documentary of the project, directed by arts faculty member Jennie Bourne, which brought a standing ovation when played at the last Faculty Assembly Meeting of the 1997–98 school year (Edelstein 1998). Other efforts to document the method and success of the project were undertaken. A student survey developed with the help of Mary Ann Sorenson, a graduate student at the City University of New York, tracked the project longitudinally and demonstrated an improvement in ecological knowledge and behavior over the grant. An extensive effort to document the method and success of the project was undertaken and detailed in a lengthy final report (Edelstein 1998).[1]

Community Impacts

The grant allowed sustainability and Ramapo to become closely linked not only internally within the college but also externally in the community. Through the Institute for Environmental Studies, a series of four annual Mid-Atlantic Environment and Sustainability Conferences were held, drawing thousands of attendees and giving the college high public recognition as a leader on this issue. I ran these conferences with a student staff, collaboration from key colleagues, and an amazing breadth of cooperation from across the college. Among the largest events ever to be held at the college, the conferences involved nearly every staff person plus scores of faculty and students. People came together with such professionalism and sincerity that they owned the success of the conferences as their own. This level of involvement was one of the most important outcomes of the conferences.

This project, which devoured my life for more than four years, became an exciting vortex of student, faculty, and staff energy. I supervised an office run by my best students, who received payment or credit to assist me. My courses became seminars focused on aspects of ecological literacy and assisting in the project. My students assisted in research needed for infusion efforts, developed and taught curriculum at the AEC, and generally demonstrated an ability to implement the project in multiple ways. My office was abuzz with activity. I spent my weeks walking the halls and discussing projects with faculty. I attended curricular meetings, administrative councils, and student club meetings everywhere in the college, giving me a sense of comprehensive understanding of my institution. The evaluation became a real research effort. The video was a new and exciting challenge. And the community outreach established contacts for transforming the direction of an entire region. This was exhausting but rewarding.

Spin-off developments of the conferences were also significant. President Scott had become increasingly interested in sustainability as a by-product of his work on global education. When several colleagues and I sat down to lunch with him at our third conference with ecological literacy gurus David Orr and Tony Cortese, among others, the conversation turned quickly toward how Scott could make Ramapo a model for sustainability. He adopted this goal with relish. Later, he brought in partners and funds for the fourth conference in the series, which principally addressed global climate change, and personally acted as the conference chair.

Given the president's keen interest, senior administrators felt safe with sustainability as a core operational construct rather than an academic topic that had little to do with how they ran their programs. Contributing to this openness was the experience of the college's Health and Safety Committee, which for many years I had cochaired with the vice president for administration and finance on behalf of the American Federation of Teachers faculty union. Committee membership included virtually the entire senior administration who played roles relating to facilities and campus life. The committee developed a consensual approach and good internal relationships while successfully tackling such issues as unsafe working conditions due to environmental causes, fire safety issues, and reduction of toxic chemicals in campus operation, including the science labs. The committee was a natural venue for diffusing ideas about sustainability to the administration and offering leadership toward sustainability for the college community. Among the growth activities undertaken through the vice president for administration and finance was sponsorship of a visit from Harvard University's campus sustainability coordinator, Leith Sharp.

Partnerships Across New Jersey

Just as the FIPSE grant ended, a further step toward sustainability was made possible by a grant from the Geraldine Dodge Foundation. With administrative support, several faculty launched a Ramapo-based multicampus project, called the New Jersey Higher Education Partnership for Sustainability (NJHEPS), for which I served as the first project director. My tenure at the helm was not an easy one. The Dodge funding was much less than we asked for. I had to fight the board to use the bulk of the money to hire an executive director rather than merely handing project moneys back to the campuses. After getting the organization properly organized to do business, I stepped down from the helm. NJHEPS has subsequently been successful in networking educational institutions across the state, promoting development of on-campus sustainability committees, offering conferences and workshops on numerous topics targeted to the needs of key campus players, and initiating an effort to make higher education a leading sector promoting greenhouse emission reductions. Under its obligations as an NJHEPS member institution, Ramapo created a sustainability task force headed by the deputy director of facilities. As a further sign of the evolving participation of campus constituencies, the first project to come from this effort was the introduction by campus police of bicycle patrols intended to promote a sense of community cohesion and to reduce inefficient vehicle trips.

Greening the Village

An emergent lesson from these experiences is that success may occur on one or many fronts while progress toward sustainability is undermined or delayed in others. As a case in point, by the end of the 1990s, there was widespread recognition at Ramapo that new construction projects on the fast-developing 300-acre campus were lagging behind the broad goals of sustainability, and opportunities to do things better were being lost. Even at the end of the millennium, buildings intended for long-term use, including two new dormitories and an arts building, were being built in a conventional manner, creating unnecessary continuing energy and resource demands. Rather than helping the campus achieve its potential as a model and leader for sustainability, these building projects merely perpetuated an unsustainable status quo. A master plan update in 2000 failed to address this shortcoming.

Sustainable architecture demands an open process of involvement in order to meet the needs of users and widen creative input. Yet Ramapo was unwilling to involve faculty with expertise in shaping key projects and reluctant to deeply involve the larger campus community in decision making. Public involvement was seen as antithetical to expeditious planning. As a result, for example, the sustainability-oriented faculty had no input into a dormitory built in 1998 and only brief input to another built in 1999. However, these new buildings demanded new infrastructure. Responding to a timely opportunity, the college took a bold and high-profile administrative decision to use fuel cells to power the new dormitories and provide uninterrupted power supply. By taking this lead with a promising new green technology, the ghost of the failed solar building seemed finally to have been exorcised.

When development began in late 2000 on the Phase VII housing project, later named "the Village," it became the first campus project to list green architecture explicitly in the request for proposals. However, these good intentions were soon sacrificed in the rush to meet an immediate demand for more campus housing. Given that the timetable for planning and implementation was impossibly short, reflecting housing demand and funding mechanisms in place, any steps that would delay the project came to be seen as threats to its success. Devolving to crisis management, it quickly became clear that there was no room in the planning process for innovation. Moreover, the Village design process continued a tradition of restricting access of community members to the planning committee. As always, the combination of haste and insularity was to prove costly.

I hope that the shortcomings of the Village planning process will become an important learning milestone for the Ramapo community. A subcommittee of the board of trustees and selected college officials formed a planning committee empowered to site the housing and get it quickly designed and built. With 300 acres, previous campus planners had always believed they had a lot of land to work with and few constraints. As a result, early planning decisions squandered the most suitable available lands. Continuing this mind-set, the Village planning committee conceptualized their project as a low-rise town house development for 525 upper-level students, mostly seniors. It reasoned that seniors, the intended residential population, would overwhelmingly want apartment-style housing (an assumption not initially borne out in reality). They further argued (incorrectly) that three-story structures would not require expensive elevators to address Americans With Disabilities Act requirements. The committee considered its town house concept to be final and beyond debate. Unwillingness to rethink the low-rise, big-footprint town house concept locked the college into avoidable and serious environmental impacts. In fact, the fast-tracked siting process became snarled in a series of environmental challenges that should have been anticipated. Legally protected but previously unmapped wetlands and water courses limited options for siting the sprawling complex. Old-growth forests occupied still another site, and at the urging of our faculty ecologist, Eric Karlin, also the dean of science, the committee voluntarily opted to protect this resource. Left with few remaining places able to accommodate the huge town house footprint and running precariously behind schedule, the committee chose a location never considered previously, atop both the AEC and a sizable area of scarce campus parking, which would have to be replaced elsewhere. Two steps back, but how to take one step forward?

As fate would have it, I was walking by the room where the committee met just seconds after this decision was made. Seeing friends in the room, I crashed the meeting. Committee members were exhausted and in shock at the trade-off they had just made. The moment allowed for a candid discussion of how the college might mitigate the loss of the AEC. The idea emerged for a new Sustainability Center to be built on a nearby parcel, a proposition that the board later approved, granting $500,000 toward the cost of design and construction.[2]

From Green Buildings to Green Campus Life

The siting delays had a further negative effect. To speed the design of the town houses, the intended green design was dropped in favor of the conventional and fast. As mitigation for this second problem, I proposed to the Ramapo College Foundation a plan to write a detailed program or blueprint for greening the social (as opposed to architectural) design of the Village. With the crucial support and commitment of Nancy Mackin, the dean of students, the proposal was funded. The project was undertaken with the help of a very competent student research associate, Kate "Ali" Higgins, and students in my spring and fall 2001 "Sustainable Communities" classes. Completion was targeted for February 1, 2002, the point at which program staffing and funding, as well as residency, decisions would have to be made for the Village.

This challenge was fascinating. Obviously, it would have been best to green the buildings themselves. But we argued that this tech-fix sustainability can be easily undermined by unsupportive residential behavior and social norms and systems. Conversely, it is possible to develop a culture in which people are enabled and motivated to live sustainably in conventional buildings. This social design may be more important than green buildings in fostering sustainable outcomes. Moreover, success in achieving sustainable goals in a conventionally built complex would have wider generalizability than would green architectural innovation. If we could green the Village, then the approach should be applicable to other existing housing units at Ramapo and perhaps elsewhere.

As with my prior experience with sustainability initiatives at Ramapo, this initiative again generated broad support. Cooperators including the Ramapo College Foundation, Deans Mackin and Karlin, Gene Dubiki (the campus architect), James Quigley (the director of NJHEPS), Bill Makofske and my other environmental studies colleagues, and the staff of residence life, facilities management, and our

Cahill Center for Experiential Learning all freely contributed to our work. Unlike the fast-tracked building design driven by costs, this was a project where people could roll up their sleeves and cooperate in a more academic manner.

The program of sustainability for the Village was intended to begin at the point the student elected to live in the new complex. A comprehensive twelve-point sustainable living pledge was designed for inclusion in the housing application. Thus, from the onset, students, by selecting to live in the Village, had to make a commitment to sustainable living. Staff hired or internally recruited to work in the Village would also have to be chosen to support, educate, and model for this mission. These included residential staff for the Village complex, for the campus store to be located there, maintenance and facilities staff concerned with supporting Village functions related to sustainability, college staff concerned with experiential learning programs, a director and staff educator for the adjacent Sustainability Center who would be directly involved in the supervision of interns, and cooperative education students and volunteers running key sustainable programs. Finally, college faculty were expected to play a key role by offering educational programming in support of the sustainability mission. These activities were related to the key indicators of sustainability, referring to measurable conditions that indicate by their achievement or absence a relative status of sustainability. The indicators initially addressed by the project are:

- *Materials cycling,* specifically recycling and composting
- *Consumption,* specifically food and purchasing
- *Efficiency,* specifically energy and water use and transportation
- *Place identification,* specifically landscape access and historical appreciation
- *Health and safety issues*

Detailed programs were developed for these indicators, clearly stating the mission, a step-by-step scenario for implementation, training and educational support information and source materials, positions required and the roles of key actors, preresidency preparations for the college and the resident, resources and facilities required, and a budget. Appropriate feedback and reward systems were developed for each indicator (Edelstein and Higgins, 2002).

For example, place identification projects seek to build understanding and respect between residents and the natural and landscaped environment surrounding the Village, as well as its cultural history. Related projects included building trails to connect the Village to specific on-campus wetland, old-growth and riparian bluff areas, and off-campus state and county parks, as well as developing educational and recreational programs using these trails and sites. Gardening projects included community gardens as well as indoor planting for food, aesthetics, and air quality. And to create active outdoor public spaces that engender involvement, recreation, mobility, learning and connection, participatory landscaping of a bird courtyard ("bird court"), a butterfly and wildflower courtyard ("butterfly court"), and an edible plant courtyard ("edible court") are contemplated. These courtyards will illustrate nonlawn permaculture yards involving native perennial or self-reseeding plants requiring little care and water, offering aromatic and colorful displays, and providing habitat creation for birds and beneficial insects. Promotion of cultural place identification would occur by using historical building and place names and issuing an artists' challenge for community art projects reflecting native American and early colonial history as well as local flora and fauna.

These sustainability programs are intended to be managed by student interns supervised by staff at the Sustainability Center and the Cahill Center (Ramapo's experiential learning program). Seven co-op positions are envisioned:

• The nature trail manager, responsible for promoting residents' contact and awareness with the surrounding natural areas of the campus and nearby parks

• The recycling coordinator, responsible for implementing the Village's comprehensive recycling program

• The compost coordinator, responsible for developing programs to collect and manage composting materials and to promote use of vermiculture

• The conservation manager, responsible for managing water and energy conservation programs

• The food and purchasing coordinator, responsible for promoting healthy and nutritious eating habits, modest consumption, and use of nonpolluting, safe, less packaged, natural, and effective products, as well as promoting participation in sustainable and local agriculture, community gardening, and food cooperatives

• The gardner in residence, responsible for the community gardening projects involving food production, landscaping courtyards, and indoor plants

• The community job and transportation coordinator, responsible for coordinating involvement of residents in community jobs, promoting carpooling, and managing a local currency system

Key support facilities for the project include the Village store and the Sustainability Center. The Village convenience store will be a source and learning center for environmentally friendly and healthy products, sold in bulk and dispersed in reusable or easily recyclable containers; a center for disseminating skills; and a site for renting recharged batteries and collecting materials for reuse and recycling. While this tiny store can offer only a limited number of products, it is envisioned as a secondary source of additional products through such approaches as a Web page linked to primary sources, outreach fairs (for example, for health food providers) and farmers' markets on campus, field trips to buy healthful, fresh, international, diverse, and local foods, and a food co-op and purchasing club employing bulk buying, food delivery, and orchestrated breakdown to the individual buyers' containers.

The Sustainability Center, intended to replace the AEC, will provide demonstration and practice facilities for the campus and community, including organic gardens, permacultural plantings, composting facilities, use of environmentally preferable products, and appropriate and alternative energy sources based on conservation and renewables, as well as other outreach and support programs addressing sustainability.

The normative system for supporting the project includes several components. The college is committed to a program of ecological maintenance and landscaping. Residents will participate in a rewards and incentive program employing a local currency, awards and recognition, and other means of social reinforcement. Sustainable norms are to be created from the outset, in the resident applicants' commitment to the mission; in the "what-to-bring" letter sent to residents before the semester begins; in orientation by the residence and project staff; through the store, a Web page, a series of interpretive prompts, and feedback tools related to performance of apartment activities; and involvement in numerous project-related activities that will help in turn to create social networks. For example, since residents might be living in apartments for the first time, the challenge of cooking opens great possibilities. Cooking classes and dinner groups, as well as the Village Store, will be used to coach students about sustainable approaches to choosing and finding ingredients, managing a kitchen, minimizing packaging, addressing recovery of organic and noncomposting wastes, and achieving healthy diets. These efforts will be rounded out by the food cooperative, food buying club, food fairs and farmers' market, and opportunities to participate in community-supported agriculture and community gardening activities.

A fifty-five-page report detailing this project was presented to the Ramapo administration in February 2002. Despite an intent to embrace the project seriously, several independent events have interceded. First, a report that habitat for the endangered wood turtle was affected by the project and its associated road and parking lot expansion projects led to a stop-work order and a protracted, acrimonious, and public dispute with the New Jersey State Department of Environmental Protection. Although eventually settled so as to allow work on the project to be completed, the project schedule was set back, and yet another environmental hurdle to future campus development was identified. As a result of the cumulative delays, only one quad of the new complex was ready for occupancy by the target of September 2002. Second, efforts to fully fund the Sustainability Center project lagged behind its construction timetable of early 2003. In addition, a serious statewide economic downturn worsened by the disasters of September 2001 inspired serious cutbacks in state funding for education. As a result, the Village and, by extension, the greening project were robbed of most staffing and support.

These events cumulatively necessitated only a modest initial implementation of the greening project. Despite a hilarious miss-start, in which confused employees marked every item on the Village Store shelves as "green," the store now features a continually increasing stock of sustainable items. Two cooperative education positions have been filled for the recycling coordinators, and with their help, the Village recycling program has been completely revamped. A vermiculture box was placed with wide fanfare into the Village apartment of the student government president, opening that organization's direct involvement with creating a sustainable campus and motivating other residents to request worm boxes. The Village plans continue to unfold and to have unexpected consequences. While a plan to implement the Butterfly Courtyard in spring 2003 was delayed pending a campus landscape plan, the situation spawned the development of an ecological landscape and arboretum concept that has thus far received positive administrative support and is likely to be at least partially implemented. So the courtyard was not completed, but a broader change was put in process. Implementation of the rest of the Village program is likely to show this same kind of delay-opportunity dynamic.

A Critical Moment

This social experiment is at a precarious juncture. Its full implementation and success will spill across the Ramapo campus. Students and staff will take the lessons home to their families and communities. And progress will be made on the crucial issue of whether the next generations of Ramapo students learn to live lightly on this earth. Through NJHEPS, Second Nature, and other networks, conferences, and publications, outcomes can be shared and then compared with lessons from other experimenting campuses. Yet as our solar fiasco in the late 1970s illustrated, failures also teach lessons, both corrective but also at times discouraging of further risk taking and innovation. For this reason, the project is a gamble that has to succeed. Although the success of the greening project for the Village rests on many actors and developments, it illustrates another key lesson of sustainability: while the success of projects demands that they be owned by participants so that they become self-driven, the work of the designer is not finished until the conditions are met for the project to be self-sustained. In this instance, we are still far from that point. The challenge now is to create the conditions and mechanisms for that transfer of responsibility to occur.

Although the Greening the Village project is the latest in a continuing saga of evolving sustainability at Ramapo College, it is hardly the end. Similar innovative programs can be envisioned for other sectors of college sustainability. For example, while the Greening of the Village focuses on the issues of campus and culture and the FIPSE project focused on curriculum, I have long sought to fund a program that will address the community as well. Building on our successful series of Mid-Atlantic Environment and Sustainability Conferences, I long ago envisioned extending the idea of the "Science Shop" to make Ramapo a

"Sustainability Shop." By this, I mean that faculty and students would work on problems—social, economic, and environmental—that bear on the quality of communities local to the college. Through NJHEPS, we have created a clearinghouse for such innovations in order to model and inspire other actors within the academic sector. Perhaps the first role of the Sustainability Shop is to create in the college or university a model of sustainability for the community to emulate.

Offsetting these challenges is the one-step-forward and two-steps-backward theme of sustainability on campus. After years of cultivating his involvement, President Scott departed for another post. After a one-year interim presidency, a new president, Rodney Smith, took office. In fall 2002, a new faculty committee to comment on campus development was created and implemented at Smith's direction. This Sustainable Buildings and Grounds Forum quickly engaged difficult issues, including the failure of the college to make post-Village projects green. And despite the uncertainties and delays of this transitional period, an important milestone was achieved in 2000 when the Ramapo board of trustees amended the college mission statement to incorporate language on sustainability. Long in the planning, it was heartening that this step had the unanimous support of the faculty assembly and easily attained board approval. Subsequently, a strategic planning initiative by the new president has generated guidelines for introducing sustainable practice to the campus. The box cites the current mission language on sustainability, a proposed revision contained in the Strategic Plan, and long-range goals, a vision, and actions toward sustainability elaborated in an entire chapter of the plan. The plan was approved by the board of trustees in spring 2003. Fortunately, sometimes the one-step forward is a giant one.

Conclusions and Observations

What does the Ramapo case tell us about the problems of sustaining sustainability, particularly during the period of a green transition from old patterns of business as usual? Our experiments in sustainability have come far enough to demonstrate some important lessons.

First, there is a threshold for sustainability. After a sufficient level of institutional experience and success, sustainable actions beget more sustainable actions, and sustainability becomes self-fulfilling. This occurs when there is movement from one or a few sustainability activists to a

Institutional Sustainability Commitments

Current mission as of March 2003[a]	Ramapo's well-credentialed faculty pride themselves on teaching and seek to educate students to become lifelong learners. They emphasize critical thinking and the awareness of value questions, such as the importance of promoting a sustainable environment. They present a challenging educational program so that Ramapo graduates can pursue truth in an increasingly interdependent and intercultural world.
Proposed mission revision as of March 2003, p. 10[b]	Ramapo College of New Jersey(s) . . . curricular emphasis include the liberal arts and sciences, social sciences, fine and performing arts and the professional programs within a residential and sustainable living and learning environment.
Strategic plan long range goals as of March 2003, p. 5[b]	By 2012, Ramapo College will be recognized nationally for its sustainability actions, facilities and programming.
Strategic Plan Vision Statement as of March 2003, p. 11[b]	As we transition to a more residential community we must broaden and expand the management of the campus to meet the needs of students in safe, healthy, sustainable and intellectually stimulating environment. We envision a campus wide re-commitment to efforts at developing and maintaining a sustainable or green environment, while recognizing that such a commitment requires dedication of planning time, inclusion in all long-range planning efforts, and the availability of adequate resources. We recognize that such efforts will require both long-term commitment and phased-in planning as part of campus-wide capital projects programming. We envision the creation and maintenance of an aesthetically pleasing and intellectually rewarding sustainable living and learning environment.

Strategic Plan as of March 2003, pp. 38–40[b]	Long Range Goal 1.6 Sustainability. Strategic Initiative 1.6 Model Sustainability Construction Ramapo College will construct a model teaching and learning facility, the Sustainability Center, to advance sustainable building and learning practices. Objective 1.6.1 *Model Sustainability Classroom/Laboratory Space* Provide a model sustainability classroom and laboratory space for Ramapo College environmental programs. Objective 1.6.2 *Sustainability Outreach* Provide a teaching/learning facility and demonstration building to serve as a resource for sustainability for the campus, K–12 levels, and the general public. Objective 1.6.3 *Sustainability Resource Center* To provide a Center as a grant funded national clearinghouse to house and distribute information and knowledge about sustainability to the College, public schools, and the general public.
Strategic Plan as of March 2003[b]	Involves the new Faculty Assembly Sustainable Building and Grounds Forum in all discussions of campus change and expansion.

[a]<http://www.ramapo.edu/about/RamapoMission.html>.

[b]*Strategic Plan, Ramapo College, 2002–2012.* <http://guide.ramapo.edu/content/StratPlan/docs/StrategicPlanFeb2003.pdf>.

much broader ownership of the mission, when concrete actions toward sustainability have been identified, understood, and seen as desirable and when these actions fit with the varied areas of responsibility, competence, and agenda setting held by multiple institutional actors. Institutionalization of a sustainability mission is the eventual outcome.

Successful institutional changes toward sustainability are indeed exercises in social learning. Institutional learning styles and capabilities vary, and it may be necessary to build the learning capacity before sustainability actions can succeed. The Ramapo history reveals that sustainable change requires unlearning conventional approaches while learning new ones. Because colleges and universities are such partitioned institutions, links between partitions are necessary to create an interdisciplinary and integrated approach. Ramapo's success has a lot to do with its being internally linked. In contrast, such learning is often defeated in many institutions by bureaucracy, departmentalization, and lack of internal institutional communication and cooperation. While hardly utopian, Ramapo College has been fertile ground for sustainable actions in large part because it is a young institution, inherently interdisciplinary, both large enough and small enough to benefit from the advantages of scale, and has had comparatively little turnover of faculty and staff over time so as to build long-term relationships, yet with a climate of experimentation and learning whereby those long-term staff have not become stagnant. The normative structure overall has tended to support cohesion rather than alienation, despite some of the scars found in academia generally. Students have historically played an elevated role in the college, giving them special influence, respect, and power and meaning that their mobilization for an issue is seriously noted. These and other idiosyncrasies of the college may be suggestive of institutional conditions that are conducive to sustainable experimentation.

The inverse of this lesson is that until the institution reaches the level of immersion in sustainability found at Ramapo, sustainability advocates need to look for footholds—fertile niches within the setting that are conducive for successful models of learning. Success in one institutional sector can inspire interest in others. They need to find allies, looking even in unlikely places. At Ramapo, such allies have been found in every corner, from the top, bottom, and middle. I think they are likely to be everywhere if one looks the right way.

The goal is to reach a critical point where the culture of the institution supports sustainability as an implicit and even explicit goal. Thus, building a sustainable campus has everything to do with culture building. Social capital, physical resources, institutional memory and image formation, the shape of key milestone events, accessible learning opportunities, and many other elements go into this acculturation for sustainability. And because administrators are responsible for real-time decisions, there is an inherent competitive disadvantage that must be overcome by mission commitment if sustainable decisions are to have a foothold. For this reason, focusing on mission is an important target.

Finally, sustainability is a collective project. It demands consensus or respectful disagreement. Top-down support is essential, but so is grassroots support, as evidenced by the broad faculty and student involvement in our projects. Key individuals make all the difference. As the examples illustrate, sustainability projects do not just occur because their time has come. Until such time as sustainable thinking and actions become socially normative, a great deal of energy will be required to nurture—or perhap force—even modest efforts into existence. While no one person can be credited with creating sustainability at Ramapo, my colleagues and I can take credit for key experiments, projects, and events. There is nothing novel about this realization. Every innovation requires innovators. And herein lies an Achilles heel of sustainable change. The tendency for projects to become associated with dominant movers and shakers creates a susceptibility for the project to falter if the leader is removed. And if a leader gets too far out in front of his or her peers on a crucial issue, as I did for awhile in advocating for sustainability to be added to the college mission statement, they may need to pull back and let others take the lead and credit for a collective victory. A lesson of sustainability that must be followed is to create cooperative ventures for secondary benefits of participation and to allow redundancy and resiliency of social systems. It is often easier to go it alone, but never better in the long run. Depth of field is key.

It will take time to see whether the early encouraging signs of our new president's commitment to sustainability are matched by implementation over time, and whether administrative and institutional support will increase, lag, or refocus. It can take years for a president to fully understand sustainability, and a major investment must be made. We are also finally experiencing a major influx of new faculty who have yet to be involved. This continual relearning of key institutional actors must be addressed if sustainability efforts are to succeed. Along with other challenges, the message may be that if we want the path to sustainability to be made of cobblestones rather than asphalt, then we have to expect bumps in the road. Learning to address these challenges patiently and creatively is a key requisite of success. In the end, it may seem like an oxymoron, but sustainability must be sustained.

Acknowledgments

I acknowledge the close collaboration of Kate "Ali" Higgins, who served as my research associate for the Greening the Village project and who, with Audrey Schwartz, served as my teaching assistant for the Sustainable Communities classes dedicated to developing the program for the Village. Thanks also go to the students in those courses who served as primary researchers; Nancy Mackin, dean of students; William Makofske and my other colleagues in the Environmental Studies program; former Ramapo president Robert Scott; James Quigley of the New Jersey Higher Education Partnership for Sustainability; and numerous other faculty, staff, and students of Ramapo College. The Ramapo College Foundation generously supported the principal project discussed here.

Notes

1. The 1998 video, *The Ecological Literacy Project*, directed by Jennie Bourne is available from the author.

2. For a time, the long-envisioned Ecological Living Facility was also part of the project, to be funded as part of the Village complex. However, the ELF was put on hold for fear that the innovations required would delay design of the entire Village project.

References

Edelstein, Michael R. 1998. "Environmental Literacy in the Undergraduate Curriculum: Final Report." Unpublished report to FIPSE. September.

Edelstein, Michael R., and Kate Higgins. 2002. "Inherit the Earth: The Sustainable Living Pioneers: A Program for Greening Phase Seven: Ramapo College's Commitment Toward Sustainable Campus Housing." Unpublished report, February 1.

Tufts University. 1990. "The Role of Universities and University Presidents in Environmental Management and Sustainable Development." Report and Declaration of the President's Conference, Tailloires, France, October 4–7.

16

Cultivating a Shared Environmental Vision at Middlebury College

Nan Jenks-Jay

Middlebury College, founded in 1880, is a private, liberal arts college with 2,200 undergraduate students. Graduate degrees are granted by the Language School and the Bread Loaf School of English. Located in central Vermont, this rural campus is situated in the village of Middlebury, surrounded by farmland and the Green Mountains and the Adirondacks. Nearby, the 1,800-acre Bread Loaf mountain campus offers courses in summer and skiing in winter.

Middlebury College has had a long tradition of environmental education and mindfulness dating back to 1965, when it established one of the first environmental studies programs at a liberal arts college. In the mid-1990s, on the threshold of its bicentennial in 2000, the college proposed even greater advancements and leadership in the area of the environment. In 1994, the president and trustees designated the environment as one of six areas of peak excellence based on existing strengths at the college. That fall, President John McCardell stated in his convocation speech, "At Middlebury we are building an academic plan that emphasizes excellence across the curriculum with special attention to the academic peaks that are the hallmark of our identity. Environmental Studies and Awareness is one of these six peaks because of the College's long tradition of being on the forefront of environmental education in both the classroom and on campus." This broad definition of the Environmental Peak, extending beyond the classroom and academic sector, challenged us to move toward a sustainable campus. This presidential declaration was recognition at the highest level that the environment was integral to Middlebury's educational mission, operational goals, and responsibility to the greater community. We have been endeavoring ever since to integrate environmental studies and sustainability in higher education through a system-wide approach. By infusing the system with environmental principles, practices, and learning, Middlebury enriches the educational process and creates a community of individuals with a shared commitment to the environment. (See the box.)

An overview of how we have tried to change Middlebury College as a system has to begin with academics, which are linked to operations and other college activities. When people ask how the infusion of environmental education and responsibility takes place on a college-wide level, my response is, "one step at a time and in every corner and outreaching arm of the college where environmentally related teaching, research and activities occur." Actually, it is a more complex story about a bold administration, multiple pathways, and creative individuals. It requires all kinds of people, regrettably too many to mention in this chapter, and their ability to come together to make a difference in the place where they learn, live and work, at Middlebury College, which provides fertile ground for such activities.

This chapter describes how a college integrated environmental considerations campus-wide through senior-level support, a cadre of collaborations, conventional planning processes, an innovation-fostering grant program, and informal networks in an atmosphere of openness to change. In doing so, it has created a shared vision and institutional core values that move toward long-term sustainability. Two individual profiles will illustrate how attention to systemic change at Middlebury has had ripple effects around the region and beyond.

A Period of Change, Taking Stock, and Senior-Level Support

As the college entered a period of high aspiration and rapid change in the mid-1990s, I was invited to join its ranks in 1997 as director of environmental affairs, a new senior administrative position with an appointment in environmental studies (ES). The director of the ES program, biologist and friend Steve Trombulak, encouraged me to consider this position while it was still being defined. He had completely revamped and revitalized the ES program in the 1980s and, with subsequent directors John Elder in English and ES and Chris Klyza in political science and ES, built a stellar academic program. I was familiar with Middlebury from serving on an external review committee for the ES program in 1988 and getting

Systemic Sustainability at Middlebury College

- The Environmental Studies Program (ES) is one of the largest majors, with forty-five to fifty graduates each May.
- ES includes forty-three affiliated faculty members from sixteen departments.
- Of the 2,200 students enrolled each year, 48 percent take at least one ES course prior to graduation.
- Over 600 alumni are employed in environmentally related positions around the globe.
- Twelve operating budgets, endowments, and grants support the "Environmental Peak."
- Environmental programming is being developed for Middlebury's programs abroad in six countries.
- An Environmental Council advises the president of the college on sustainable campus matters and recommends policy.
- The college composts 75 percent of food waste from its dining operations
- On average, 60 percent of the solid waste generated by the campus is diverted for reuse and recycling.
- When the old science center was recently deconstructed, 1,354 tons of material (97.4 percent) was recycled.
- College forestlands are managed under Forest Stewardship Council's green certification standards.
- Energy is addressed through conservation, co generation, new technology, efficient design, and a thermal comfort policy.
- Alternative energy vehicles are leased as part of the college fleet.
- Due diligence through compliance audits ensures that the college is not violating environmental regulations.
- Regular campus environmental assessment track progress and identify areas that require improvement.
- Sustainable design and environmental building standards are employed in all new construction and renovations.
- A career services employee was named Northeast Region Environmental Career Counselor.
- Renowned environmental author Bill McKibben was recently appointed to the new position as environmental scholar in residence.
- A greenhouse gas inventory has been conducted, and plans to become a carbon-neutral campus are being developed.
- Received U.S. EPA Environmental Merit Award for institutional leadership and environmental committment.
- Received Vermont Governor's Award for Environmental Excellence and Pollution Prevention.

to know the program's directors over a twenty-year period through the Northeastern Environmental Studies (NEES) meetings, where much information was shared. I knew that Middlebury was one of the most diverse and rigorous interdisciplinary environmental programs in the country and that the teaching and scholarship were exemplary. The college had established successful campus programs in recycling and energy efficiency well before other institutions. Executive Vice President for Academic Affairs and Provost (provost) Ron Liebowitz, the person to whom this new position would report, prevailed on me to come to Middlebury from the progressive West Coast and to leave an endowed chair as director of environmental studies I held at the University of Redlands in California. As the provost and I crafted the final details of the new environmental affairs director position, my imagination was captured by the idea of implementing institutional-scale transformation on behalf of the environment.

In part, I viewed advancing the Environmental Peak, including the academic and nonacademic sectors, and evolving into a more sustainable system as a chief goal. As director, I would also chair the Environmental Council, a committee that advises the president about the sustainable campus and recommends policy, and attend quarterly trustee meetings in addition to other responsibilities. The director of environmental affairs works hand in glove with the rotating directors of the ES program, who report to the dean of faculty. Between the two positions, there is much access to the senior leadership of the college. This new appointment turned out to be as much about vision and innovation as it was about facilitation and collaboration, all being essential components of an idea born in a period of rapid change and that involved the entire institution.

Even in a receptive atmosphere where change and the environment are considered positive, it was useful to take time to assess things, especially if I were to be an agent of institutional change. Higher education is nothing more than a system—a system that is highly stratified and compartmentalized with boundaries and barriers. It is a place where patterns of behavior become comfortable and communication can be poor. Institutional memory is long, and forgiveness is sometimes in short supply; one mistake could prove to be a lasting detriment. In a system that is this complex and rigid, it is best to gain some understanding about how it really works and what is truly valued. My thought was not to create a lofty self-proclaimed vision, but to work with others to frame a realistic vision for the near future and develop pathways to achieve it, pathways that would lead to more environmentally enlightened individuals in an integrated campus.

I took a semester to understand people's motivations and to study this particular institution's culture. Middlebury College's trustees and senior administrators were not unlike the sponsors, captain, first mate, and navigator of a large sailboat. They had charted a new course, and the wind was filling the billowing sails. Through thoughtful dialogues, transparency, communication, and workshops, the institution began moving forward as a whole during this period of great change. Through a capital campaign, new resources were being widely distributed to benefit all. As a newcomer, I could sense the momentum and high morale. The anchor of resistance that comes with all things that are different had definitely been raised.

I learned that the college planned a ten-year expansion that would increase the number of students, staff, and faculty while also adding new facilities that included a science center, library, and residential halls. A new student residential commons system was being implemented. Middlebury's status was shifting from a very good college in New England to one that was highly competitive nationally, attracting a new generation of talented students and junior faculty. Its endowment paled beside the endowments at schools with which it was compared, but the $200 million capital campaign was increasing its resources. The senior faculty members were energized. The experienced administration and board of trustees were not averse to calculated risks. This college's culture was one that struck a good balance between respecting tradition and encouraging progress.

I found that exemplary programs were already in place, which a campus environmental audit conducted by the Environmental Council verified with measurable results. Middlebury's early commitment to the environment clearly laid the foundation for all that followed, but it had taken the tenacity of the ES program directors, a flood of interest from students, ingenuity in facilities management staff, and the savvy of some administrators to make it occur. Admirable as they are, these things alone working independently cannot lead to the sustained and systemic environmental commitment we were now striving to achieve—something that requires an endorsement from the highest level, the involvement of a great number of people from all across campus and their combined energy. With administrative support being key to long-lasting success, this administration's philosophical and financial support is worth noting. The former vice president treasurer formed the Energy Committee, which evolved into the Environmental Council; the president designated the Environmental Peak of Excellence and funded the Campus Environmental Grants; the vice president for academic affairs/provost created the director of environmental affairs position and operating budgets; the dean of faculty supported new shared faculty appointments in the ES Program; the vice president for facilities planning cochaired a process to develop sustainable design and construction standards endorsed by the college's trustees, and the current vice president and treasurer cochairs the Carbon Reduction Initiative Committee with me. However, involvement of a great number of people across campus working together for a common purpose had yet to occur. This was a goal I viewed as essential to promoting institutional change and creating a community vision, and so addressed it early on in the process.

A Cadre of Collaborations to Increase Participation

Not everyone considers himself or herself to be an environmentalist, even in Vermont, but many people have overlapping interests or share core values about a quality of life that includes a healthy environment. Holding this philosophy to be true, I met with individuals in all the major sectors during the first year to discuss how our shared interests could benefit each other and ultimately the college. I contended that successful environmental programs attract bright students, appeal to donors, and prepare graduates for a wide range of occupations. These concepts resonated with Admissions, College Advancement, and Career Services as we discussed ways to collaborate. Some were perplexed, others understood, but all listened—perhaps because the environment was a peak, but also because the ideas were compelling. In this period, I invested in developing personal and professional relationships. Although this process was time-consuming, the contacts needed to be constantly reinforced in order to build something far greater together over time.

These early conversations first led the vice president for college advancement to collaborate by cosponsoring a conference with the ES

Program entitled "Something Wild, Something Managed: Wilderness in the Northeast Landscape," a program not originally on the two-year schedule of events for the College's Bicentennial Celebration. It attracted large audiences and was credited for inspiring subsequent forums on wilderness in the Northeast. Next, College Advancement, Admissions, and Environmental Affairs collectively designed an Environmental Peak brochure that for the first time included sustainable campus programs intermingled with academic programs. Initially, people were uncertain about this attempt to reach different audiences with one voice, but with input from all parties involved, a high-quality publication was produced and is still used by all three offices. In addition, an environmental fact sheet describing the academic and campus programs was developed and is updated annually and provided to these offices and others, so their knowledge about the environment at Middlebury is complete and current.

On a different front, I offered to assist in the capital campaign, explaining that the environment appeals to a wide range of people from corporate executives to grassroots activists. Over a twenty-four-month period, I met with eighteen major prospective donors, three foundations, and several alumni gatherings. I realized that the alumni not only wanted to know more about environmental courses and research, but were extremely interested to learn what the college was doing to improve the environment on campus and in the region. Engagement with prospective donors resulted in over $1 million in endowments for the Environmental Peak and gifts to other campaign target areas as well, serving the college overall.

When I arrived five years ago, the Career Services Office's (CSO) definition of environmental occupations was twenty years out of date. After several starts with CSO directors who departed, the current director, Jaye Rosenborough, was hired, and she became enthusiastically supportive. When CSO staff member Krista Siringo joined the Environmental Council, I encouraged her to attend several national environmental conferences. A quick learner, she was recognized in 2001 with the Northeast Region Environmental Career Counselor Award. CSO now offers a full range of environmental job and internship listings, recruiting with prospective employers and advising. What is more, student projects related to the sustainable campus build impressive resumés and lead to post college jobs. For example, Lara Dumond, class of 2000, designed a restored wetland on campus with a National Wildlife Federation Campus Ecology grant and a Middlebury environmental campus grant. Immediately following graduation, she was hired as a wetland specialist by a Boston environmental consulting firm, whose principal was a Middlebury alumna. Links with CSO continue as several students receive internship awards each summer to study the environment. A student, Gabe Epperson, class of 2003, inspired by an ES seminar course project on campus transportation issues, received a self-designed internship from CSO focused on reducing dependence on single-occupancy vehicles at Middlebury. The college immediately implemented his shuttle bus recommendations, and the course project report is guiding pedestrian campus discussions within the administration. He was hired by Utah planning commission in his home state.

The integration of the academic sector with sustainable campus initiatives is essential, and we worked to reduce some skepticism about the academic value of campus and regional projects. Three joint appointments in the ES Program already existed in biology, English, and political science, but new shared appointments in religion, geology, history, and economics were approved through the efforts of the ES Steering Committee and were filled with junior faculty whose research is grounded in the environment. Many junior faculty see connections between sustainable campus efforts and their areas of interest or research involving energy, consumption, agriculture, and economic offsets. However, some senior faculty still viewed student projects involving the campus or region as not being serious work. There needed to be assurance that projects receiving credit were rigorous. Therefore, a distinction was made to clarify projects that are volunteer efforts, those that are paid positions, and those in which students receive course credit. In these latter instances, we raised the bar. A science professor, once a great skeptic, has subsequently advised two student research studies involving compost and wind power for Middlebury.

These steps helped more faculty see Middlebury as a sustainable system, and today most are fully committed to the environment. Partners were acknowledged for their contributions through the college's staff-faculty newsletter in a bimonthly column on the environment, in newspaper and magazine articles, in annual reports to the senior administration, and directly through letters of gratitude. Venturing forth with this experiment in institutional transformation, trying to be inclusive and integrative and respect existing areas of specialty, expertise, and territory, is a balancing act. I began to identify the many excellent class projects and impressive staff initiatives quietly taking place in pockets across campus that needed to be linked to each other and to the whole to have greater impact.

Making these connections across campus sectors and into the community is a major role of campus sustainability coordinator Connie Leach Bisson, who also provides education, information, and resources to all parts of the college; conducts outreach; and oversees new programs. Initially a recycling coordinator position, the responsibilities grew and were expanded under the leadership of Amy Seif as environmental coordinator, from which the current position emerged. This is where the real magic happens: enriching projects and building alliances into a network. A recent example started with a physics student's proposed senior thesis to study wind atop the science building. Connie knew that economics students analyzed the financial feasibility of wind generation at the college's ski area, so suggested it as an alternative site to benefit not only the student as a learning experience, but also the college by providing data about wind potential. At a meeting organized by Environmental Affairs with professors, students, administrators, facilities management staff, and the manager of the ski area, not one objection was raised, and all comments were constructive.

This positive response might not have occurred a few years earlier, but the relationships that had developed among those seated around the table enabled everyone to explore this new idea enthusiastically. The 30-meter meteorological tower cost more than rooftop equipment, but is jointly funded by an environmental campus grant, the Physics Department and Environmental Affairs. Even the permitting process went smoothly. The study will now continue for three years and involve a far greater number of students, faculty, and a number of staff who are volunteering their time. A wind power company is interested in the study, as is the State of Vermont, which has funding for wind turbines at ski areas. It is too early to say if the college will pursue installing an energy turbine eventually, but the data gathered by students will contribute to the discussion. The project will be reassessed each year, and the campus sustainability coordinator will provide continuity and communication among the individuals involved.

The campus sustainability coordinator is also working with the new residential commons, designed to be a seamless learning experience, to introduce students to campus environmental goals and initiate new programs such as a model environmental dorm room, convenient switches to reduce heat when rooms are unoccupied, clothes drying racks, bike storage rooms, energy reduction competition, and cohosted educational programs.

Initially, even simple collaborations required time and patience to educate people about achieving shared goals and the increasing value of outcomes. Now, five years later, partnerships come together quickly, but will always require enormous attention to detail and to the needs and goals of the parties involved. Independence at Middlebury has not disappeared, but has shifted toward more interdependence, and campus environmental collaborations are now widely recognized as being positive, rewarding experiences. This confidence and growing recognition has led to the ease with which complex projects like the wind study come together today.

A Combination of Conventional Methods, Innovation, and Informal Networks

Developing a strategic plan and implementation process is a conventional method in higher education to provide direction, but also to establish credibility in a system that respects such processes. However, I also wanted to encourage entrepreneurial thinking and wanted to foster informal networks as well. The structures developed from 1999 to the present have addressed all three goals. These efforts systematized some of the informal collaborations or sought to extend the collaborations just described.

In 1999, a committee appointed by the provost met to develop a comprehensive five-year plan for the Environmental Peak. Portions of previous reports were carried forward into this process, including the *ES Peak Report* by Steve Trombulak and *Pathways to a Green Campus* by the Environmental Council chaired by Steven Rockefeller, which were connected efforts but were not integrated. A highly participatory planning process chaired by the director of Environmental Affairs initially gathered data that informed the discussions of a half-dozen subcommittees. A year-long process in 1999–2000 engaged a wide range of individuals from the college community, obtaining diverse perspectives and building consensus. This process culminated in the *Environmental Peak Report and Recommendations 2001: A Vision for the Future*, with an action chart targeting all sectors of the college. The report serves as a guide for action steps and for the allocation of limited resources. It identifies all those who need to be involved and describes what is required for implementation. Recommendations are prioritized at a day-long retreat hosted by Environmental Affairs held each May for environmentally affiliated faculty and staff.

Because the environment is a fast-changing field, this plan also allows for new opportunities that arise and could not have been predicted when the report was prepared. With this in mind, the Environmental Council established a campus environmental grants program funded by the president's office as an incentive to encourage environmental innovation and leadership at all levels. This program is an incubator for new ideas, both small and large, that can be tested. The grants support pragmatic proposals to improve the campus environment, but also provide the freedom to try something new and succeed or fail without recourse. I felt it was important to recommend that proposed projects be collaborative, involving some combination of students, staff, or faculty. The process has been revised twice to be less intimidating to those unfamiliar with proposal writing and to include progress reports to stay on schedule and enable the Environmental Council to assist if recipients encounter problems. A subcommittee of the Environmental Council carried out an assessment after two years to determine if the grants were achieving intended goals and to identify unanticipated outcomes. Benefits that had not been anticipated included applicants' learning to make clear and convincing cases for campus sustainability, an expanded number of people from different areas involved in campus efforts, an increased sense of institutional pride, and the formation of new partnerships.

The campus environmental grants transcend traditional boundaries, involve many dimensions of the college, and have created new collaborations. During the past three years, twenty-six campus environmental grants have been awarded. Examples include a bicycle trailer for campus mail delivery; production of biodiesel fuel from dining hall waste vegetable oil (partially funded by the National Wildlife Federation Campus Ecology Program); a study of water efficiency at the college's golf course; and an expanded yellow bikes program making bicycles available to students on campus, and wind monitoring at the ski area. A student-staff grant testing double-sided printing in computer labs led Information Technology Services to program all public printers to produce double-sided copies by default, thereby saving reams of paper.

The Environmental Peak agenda has advanced more rapidly not only due to a traditional plan of action and innovation, but from the informal networks established along the way, forming a kind of social capital for environmental education and campus sustainability. Even in the times of financial downturn, collaborative efforts make the most of shared resources and ingenuity at Middlebury College where individuals have a can-do attitude about working together. These important intersections create a nexus where relationships and trust develop and strength and support emerge. Informal networks generate greater interest through extended conversations, establish bonds through mutual interests, and build trust through successful ventures.

Profiles about People and the Place Where They Work, Live, and Learn

Two profiles illustrate how people are making a difference at Middlebury College that have had far-reaching impacts. These stories reinforce the strength of connections and demonstrate leadership at all levels.

Food for Thought

Food is part of a cycle at Middlebury College. A decade ago with a grant from the State of Vermont, Assistant Director of Facilities Management Norm Cushman, who oversees the college's impressive recycling program, started a small pilot composting project. Today, composting at Middlebury diverts 300 tons of dining waste from landfills annually. The program's expansion demonstrates what a small amount of external funding and a good idea can achieve. Taking things a step further, a collaboration between Dining Services, Facilities Management, ES majors, and the Environmental Council was recently formed. Compost was used as a nutrient-rich soil amendment for a student-designed greenhouse in which students studied cold-tolerant greens, such as kale and hardy lettuce varieties. At the suggestion of Matt Biette, associate director of Dining Services, the greens were served regularly in the dining halls. They then made their way back into the greenhouse soil after being composted as food waste, to begin the cycle again. Another student research project studied the heat generated by the compost piles. The results not only produced a professional publication, but also helped justify a new greenhouse for dining, funded by an environmental campus grant. The soil is heated indirectly by the composting process, and organic greens and herbs are grown during the depth of the winter, when local greens are unavailable.

Otherwise, Middlebury is committed to purchasing locally produced foods and is the only academic member of Vermont Fresh Network, a partnership of restaurants and stores that support Vermont agriculture. Charlie Sargent purchases the food for Middlebury's Dining Services and obtains about one-third from local sources, much coming directly from the producers, such as Monument Farms milk, Cabot cheese, Ben and Jerry's ice cream, meats, breads, maple syrup, and fruits and vegetables in season. Stipulated in the contract with the larger food suppliers is a requirement for locally sourced products. He takes stock in the fact that the director of dining supports local food procurement, as does the president of the college. Dinners at the president's house often serve up local fare such as lamb and fiddlehead ferns in the spring. The students in an ES senior seminar traced local foods back to farms illustrated with GIS maps. Dozens of students and faculty members have been involved in various phases of ongoing dining research collaborations, and dining staff members have thoughtfully suggested purchasing products that can be composted, instead of plastic, and made other environmentally conscious recommendations. When Charlie served on the Environmental Council, he initiated an annual Vermont Harvest dinner in student dining halls with components of the entire meal coming from nearby farms and labeled as such. A native Vermonter, Charlie says he feels good knowing the college is helping to keep land open and in active agriculture.

For the past twenty-seven years, Charlie Sargent has enjoyed working for Middlebury Dining Services almost as much as he enjoys fishing. When he stepped off the Environmental Council after several terms he said, "Being involved with the council has been one of the high points of my employment at Middlebury College." That meant a lot to all those who worked closely with him on the Environmental Council and will continue to do so through informal networks. Dining Services at Middlebury College is not only growing its own organic greens during the colder months, composting food waste, and serving as a source for student research projects, but it is revitalizing rural communities by building a strong local food economy in Vermont.

Sustainable Design and Construction: Influencing Professionals and Leveraging Change

When the new science center, Bicentennial Hall (Bi Hall), was proposed in the early 1990s, the college's forward-thinking trustees identified green design and construction as a top priority for the $47 million project. This award-winning building was to have an impact on all future construction on campus, on its professional consultants, and on the forest products industry in Vermont. Following a design charette and with the trustees' directive, the architects, engineers, and project manager incorporated sustainable design, construction methods, and materials into the plan, though they had no previous experience in green building. The plan included energy efficiency through design, triple-glazed/high-E windows, insulation, fluorescent lighting, and heat recovery loop in the air system. Construction waste was recycled. Solar-powered lights illuminate the parking area. Linoleum flooring was installed, as were products made from recycled materials. Dozens of other green design features and materials were employed in the six-story building. When commissioned after it opened in 1999, College Facilities Management adjusted the energy systems to operate even more efficiently.

We learned many lessons from this project. We realized that the college was ahead of others in its desire for sustainable building design, and our consultants, whose experience was with science centers, libraries, and residential halls, not green architecture, realized this as well. As a client we were asking for something our consultants were not yet prepared to deliver. I realized that green architecture was growing rapidly and predicted that this would not be the case in five years, but until then, the college would be building and renovating at an unprecedented rate. Therefore, the Environmental Council recommended that the college develop its own sustainable design standards. In May 1999, the trustees endorsed a resolution approving guiding principles for sustainable design and construction "that embodied the philosophy and spirit of the College and . . . outlined the College's environmental goals pertaining to construction, renovation, operation and maintenance of campus facilities." Subsequently, a set of standards and framework for implementation were developed by a diverse committee and with assistance from Dan Arons, a consulting architect. The Student Government Association appointed a student to the committee as well. Middlebury's sustainable design and construction standards were put into practice and applied to every project after Bi Hall, including a residence hall that will feature a green roof, a naturally designed cooling system, and ecological landscape among other features. Designers of the new library, just breaking ground, not only seek to meet Middlebury's own sustainable design standards but are following the Leadership in Energy and Environmental Design (LEED) rating with the U.S. Green Building Council as well. Middlebury's sustainable design standards, which include LEED, will become part of the college's master plan given to all consultants. These standards have admittedly influenced each professional consultant working with the college in the past four years, who agree that they have transferred learning and sustainable concepts from Middlebury to other clients' projects worldwide.

Bi Hall also leveraged significant change in the region. Over 30 percent of the building materials came from Vermont including nearly 125,000 board feet of green-certified wood in partnership with Vermont Family Forest (VFF), a collection of small forest lot owners using sustainable practices. Due to the large volume and the college's request that it be certified and locally harvested and processed, Middlebury jump-started a new sustainable wood industry in Vermont. The building showcased six species of natural wood panel and trim, with each floor having a distinct look and warm glow. Thirty-two local businesses were involved in the process, and 75 percent of the wood came from forests within thirty miles of campus. Everyone was pleased with the appearance and proud to have been part of the project, from foresters and carpenters to the architect and trustees. Logger Bill Torrey who toured Bi Hall commented, "I got a kick out of the fact that I could trace the boards back to the woods they came from. It snapped my garter."

The project manager pointed out that "people working on the project were initially taken aback, but by the end realized that it takes a better eye, more creativity and a higher level of craftsmanship, so came away with a real sense of pride in what they had done." The forest products industry is slow to change, but according to David Brynn, the Addison County forester and founder of VFF, "The Bi Hall effort opened new channels for the manufacturing of wood products that can be certified as coming from sustainable managed forests in Vermont." Because the process was uncharted and arduous at times, Environmental Affairs cooperated with VFF to produce a report tracing each step and identifying how to make the process more efficient and locally grounded. The report guided many changes in this fledgling industry and gave others the confidence to participate. Middlebury was not only in the forefront of green design on college campuses; it became a capacity builder for a sustainable wood industry in Vermont. I recall the expression on Vice President for Facilities Planning David Ginevan's face when I said that Middlebury was not merely purchasing wood products, it was facilitating economic and environmental change in the region. He clearly saw the potential of what more Middlebury could do in Vermont and began working with the newly formed Cornerstone project, a group of large businesses dedicated to matching markets with new suppliers of local and Vermont green products. I helped to found Cornerstone, and Middlebury became a role model for other businesses. We then decided to put our own forestland into green certification and include permanent research plots for faculty and student studies.

When the architects for Ross Commons, a proposed student residence, specified clear cherry from Pennsylvania, we requested that they consider wood from Vermont forests. It took some strong encouragement from myself and the members of the Project Review Committee, but samples of character wood panels from VFF-certified forests were taken back for review. Upon examination, one architect became so enthralled with the design in the character wood for birch that the specifications were changed from cherry to Vermont-certified beech, birch, and sugar maple, 62 percent of which came from Middlebury's own forestlands. Flooring, ceiling, and wall panels and furniture made from Vermont green-certified wood adorn this new building. The architect later wrote to VFF, "Your fundamental concern for managed use of and replenishment of our forests is admirable. More Americans need to be so involved, particularly those of us who've been rather indiscriminate in our usage. Education is the key. You are teaching us that managed sustainable forest land is not only necessary, but economical." What a shift from our initial meeting. Having this kind of influence on a renowned architectural firm working internationally was truly gratifying.

While Bi Hall used 30 percent locally sourced materials, this project had 50 percent of the materials sourced from Vermont and the rest of the Northeast. When the dorm opened this fall, each resident received a fact sheet describing the environmental features and suggesting ways to reside there more sustainably. At the same time that Ross Commons was constructed, a new state-of-the-art recycling center also utilized wood from Middlebury-certified forestland. Two student projects associated with Ross Commons and the recycling center resulted in an article in *Northern Woodlands* magazine and an exhibit documenting the wood from forest to installation. In 2003, the college received an Environmental Merit Award from the U.S. Environmental Protection Agency for institutional leadership and commitment to creating sustainable communities for its role as a catalyst in promoting the use of green certified wood.

Vermont green-certified wood will be used for the next student residence and the new library. Through these projects, we have learned that by going directly to the forest or to producers and not to a lumberyard or catalogue for products, as is done traditionally, we have to start much earlier and be more thoughtful about the process to meet completion deadlines. Middlebury has also found that through its imperative for sustainable design and construction, it can influence professionals, from local carpenters to international architects, about the importance of local economies and environmental quality as it also leverages large-scale change in the region.

The Right Direction

As Middlebury College entered a period of rapid change, it created a direction that reflects the institution's overarching academic and institutional values—values in which the environment is integral. By integrating environmental considerations into its planning processes, Middlebury is better prepared to assess risks, identify opportunities, and make more informed decisions. By establishing key administrative, faculty, and staff positions and the Environmental Council and by preparing strategic plans and environmental standards, we have institutionalized a commitment to environmental integrity on campus. By encouraging creativity and stimulating progressive thinking, new ideas have been generated; faculty, staff, and administrators are invigorated; and students are inspired. By realizing its role and responsibility in a broader community, Middlebury has leveraged greater economic and environmental sustainability within the region and influenced professionals having far-reaching impact.

Higher education should be a participant in the current dialogue that is taking place about how to define and achieve sustainability. The academic sector needs to gain a clearer understanding of our relationship and responsibility to these critical global issues. As we proceed with our efforts, it should be in consort with others taking our work to a greater scale. While Middlebury College strives to lead intellectually and pragmatically with a dynamic model for systemic learning and advancement, we realize that there are no perfect models for becoming a sustainable society. Instead, perhaps there is just a direction—a direction that Middlebury College as a community is consciously taking as a pathway to the future.

About the Authors

Peggy F. Barlett is professor in the Department of Anthropology at Emory University. A cultural anthropologist specializing in agricultural systems and sustainable development, she carried out fieldwork in economic anthropology in Ecuador, Costa Rica, and rural Georgia (U.S.A.). Earlier work focused on farmer decision making, rural social change, and industrial agriculture. Recently, interests in the challenge of sustainability in urban Atlanta have given her an opportunity to return to early training in applied anthropology and to combine it with interests in political economy, group dynamics, and personal development. Currently, her sustainability work at Emory University has drawn her to "play in the creek" with her local Watershed Alliance and to explore the power of connection to place in fostering change within the university and Atlanta neighborhoods. <http://www.emory.edu/COLLEGE/ANTHROPOLOGY/FACULTY/ANTPFB/index.html>

Richard Bowden associate professor in environmental science at Allegheny College, has liberal arts undergraduate training in biology and environmental studies and graduate training in forest ecology and biogeochemistry <http://merlin.alleg.edu/employee/r/rbowden/>. He collaborates at several National Science Foundation Long-Term Ecological Research Sites, studying natural and human disturbance to forest ecosystems. His interest in applying academic knowledge to environmental problem solving was fostered during outdoor activities as a Boy Scout, underpinned by a relationship between stewardship and his Catholic faith, and driven by student desires to "do something useful" with academic knowledge. His courses engage students in hands-on projects (e.g., stream protection, forest management, urban restoration) relevant to the local and regional communities. Bewildering many students, he is an avid hunter, spending much time afield appreciating nature's richness and complexity. Some of his best reflection has come while sitting in a tree. He and his family enjoy cross-country skiing, fishing in northern Minnesota, and kayaking local lakes and streams.

Geoffrey W. Chase has been the dean of undergraduate studies at San Diego State University since January 2002, following nine years at Northern Arizona University. Prior to that, he was a member of the faculty in the School of Interdisciplinary Studies at Miami University of Ohio for eleven years. His chief professional interests are undergraduate curriculum, interdisciplinary education, and the evolution of higher education from teaching to learning institutions. He finds comfort and exhilaration in many different environments: a frozen lake at midnight in Finland; above tree line in the San Francisco Peaks in northern Arizona; in the rhythm and swell of surf off Pacific Beach, California; and gazing out over Buzzard's Bay from the top of the Gay Head cliffs at the southwestern tip of Martha's Vineyard.

Audrey B. Chang received an M.S.E. in energy engineering and a B.S. in earth systems from Stanford University. In recognition of her passion for environmental activism and education, she was honored with the university's 2001 Lloyd W. Dinkelspiel Award. Her interests center around energy and environmental issues, including efficiency, renewable energy, sustainable development, and green buildings. She enjoys spending time outdoors and sharing her enthusiasm with youth through environmental education. Currently, she works as a project manager for Energy Solutions, an energy efficiency and environmental consulting and project management company in Oakland, California. Continuing her efforts to help bring sustainability to Stanford, she is managing the Leadership in Energy and Environmental Design certification of the Sun Field Station at Stanford's Jasper Ridge Biological Preserve.

Bruce Coull is dean of the School of the Environment and a Carolina Distinguished Professor of Marine and Biological Sciences at the University of South Carolina (USC). Following work on small sediment-dwelling animals in Bermuda for his Ph.D. from Lehigh University, he was an oceanographic postdoctoral fellow at Duke University and then faculty at Clark University prior to joining the USC faculty in 1973. He was a senior Fulbright scholar at Victoria University of Wellington, New Zealand, and a visiting professor at the University of Queensland, Australia. The author of over 125 scientific papers in marine ecology, he is now particularly interested in educating students about how to sustain the earth. (The frustration is that this is not occurring fast enough.) Solace and reflection come during "ecosystem sampling expeditions," particularly in those incredibly productive coastal salt marshes and freshwaters of the Southeast, where an elusive Pisces can be lured and subsequently released.

Laura B. DeLind is a senior academic specialist in the Department of Anthropology at Michigan State University. She has spent many years working in her own midwestern backyard studying patterns of everyday life and the impacts of daily choices and behaviors on our environment, communities, culture, and sense of ourselves. As a scholar-activist, she writes about the contemporary agrifood system, paying particular attention to the long-term costs and inequities at the local level. She is an advocate of more placed-based and democratized systems of food production, distribution, and consumption and of waste management, both on and off campus. In 1996, she established Growing in Place Community Farm, a working-member community-supported agriculture project in Mason, Michigan. In 2002, her university Committee on Campus Food and Agriculture began working with food stores and dining hall services to study the campus food system and to explore more environmentally and socially sustainable alternatives.

Michael R. Edelstein is professor of environmental psychology at Ramapo College of New Jersey. Since 1979, he has conducted research on the psychosocial impacts of environmental contamination. He is the author of *Contaminated*

Communities: The Social and Psychological Impacts of Residential Toxic Exposure and (with William Makofske) *Radon's Deadly Daughters: Science, Environmental Policy and the Politics of Risk*, as well as coeditor with Makofske of *Radon and the Environment*. His research has also involved the impacts of environmental change on indigenous peoples and the practice of environmental impact assessment. Over the past several years, he has become involved in environmental exchanges with Russia, serving as a project director for two recent grants from the Trust for Mutual Understanding. He has worked extensively on greening college campuses and served as the first project director for the New Jersey Higher Education Partnership for Sustainability, a consortium of campuses, supported by the Dodge Foundation, seeking to make their operations, curriculum, campus life, and community relations more sustainable.

Alan W. Elzerman directs the School of the Environment at Clemson University, where he is also chair of the Department of Environmental Engineering and Science. After completing a Ph.D. in the water chemistry program at the University of Wisconsin–Madison, he came to Clemson in large part because the topography reminded him of his native Massachusetts. He has maintained this connection to the land ever since. His primary teaching and research interests are environmental and analytical chemistry. His research efforts have been directed toward the sources, fate, and control of chemicals in the environment, as well as toward improving analytical techniques. He has led research efforts on polychlorinated biphenyls (PCBs) and polycyclic aromatic hydrocarbons (PAHs) in the lake that adjoins the Clemson campus and has used his expertise to assist local community groups interested in the environment.

Paul Faulstich is associate professor and chair of the Department of Environmental Studies at Pitzer College, one of the Claremont Colleges. A human ecologist by training, his professional interests include the ecological dimensions of human ideologies, the ecology of expressive culture, and questions of how—not just what—nature means to people. In 2002–03, he was a senior Fulbright fellow and visiting scholar at the Centre for Resource and Environmental Studies at the Australian National University, pursuing a project called "The Natural History of Place-Making." His field research is primarily in the red sandy desert of Central Australia, working with Warlpiri Aborigines. His greatest loves are his two daughters and the slickrock country of the American Southwest.

Allen Franz is a professor at Marymount College, Palos Verdes, where he teaches courses in anthropology, ecology, geography, ethnic studies, and interdisciplinary studies; he also coaches soccer. As an anthropologist he has conducted research in Mexico, Spain, and the American West and Southwest on cultural ecology and geography; regional and ethnic histories (particularly Zacatecas and the Huichol Indians); farming practices, agricultural labor and labor organization; and open space management and land and habitat restoration. He is active in local and regional organizations addressing educational and open space and habitat management. In his spare time, he works on his house and yard and enjoys music, sports, hiking, cycling, kayaking, and astronomy.

R. Given Harper is professor and chair of biology, and associate director of environmental studies at Illinois Wesleyan University. He teaches courses in ecology, environmental issues, and experimental design, in addition to a "Tropical Ecology" course in Costa Rica. His primary area of research is avian ecology, and he has focused most recently on documenting organochlorine pesticides (e.g., DDT) in birds and other wildlife.

Abigail R. Jahiel is associate professor of environmental and international studies and director of the Environmental Studies Program at Illinois Wesleyan University. She teaches courses on environment and society, American environmental history, and international and comparative environmental politics. As well, she offers experiential learning seminars and teaches Chinese politics. Her primary areas of research have been the study of environmental politics and policy in China and the effects of globalization on environmental issues there.

Nan Jenks-Jay has been a researcher, professor, and administrator involved in environmental education for over twenty years, from community college and liberal arts colleges to university programs. She has been associated with the two oldest undergraduate environmental studies programs in the country at Williams College in Massachusetts for fifteen years and with Middlebury College in Vermont. Between them, she developed new undergraduate and graduate environmental programs for the University of Redlands in California, where as director of Environmental Studies she held the Hedco Endowed Chair. Returning to New England to accept a new position as director of environmental affairs with Middlebury College, she has created an integrated vision for the future and provides leadership in advancing the college's environmental academic program and sustainable campus. She has written extensively and speaks throughout the United States and in other countries.

Patricia L. Jerman has degrees in biology from Bucknell University and public administration from the University of South Carolina. She has managed the Sustainable Universities Initiative since its inception in 1997. Prior to coming to the university, she was executive director of the South Carolina Wildlife Federation, a consultant for a Fortune 500 engineering and environmental firm, and environmental adviser to former governor Dick Riley. Her environmental interests were shaped by her parents, for whom composting and following changes in nature through the seasons are second nature.

Robert S. Lawrence is the Edyth H. Schoenrich Professor of Preventive Medicine and associate dean at the Johns Hopkins Bloomberg School of Public Health and is the founding director of the Johns Hopkins Center for a Livable Future. A graduate of Harvard College and Harvard Medical School, he has worked in community health at the University of North Carolina, as the first director of the Division of Primary Care at Harvard Medical School, and as chief of medicine at the Cambridge Hospital. He chaired the first U.S. Preventive Services Task Force from 1984 to 1989. Equity and justice are guiding principles for his work. A founding board member of Physicians for Human Rights and president from 1998 to 2002, he has participated in human rights investigations in eight countries. In October 2002, he received the Albert Schweitzer Prize for Humanitarianism for his lifelong effort to improve health care, human rights, and the environment.

Terry Link is director of the Office of Campus Sustainability at Michigan State University and an academic librarian for more than twenty-five years. He initiated a campus movement that resulted in a university committee and ultimately his office, initially funded through a grant from U.S. Environmental Protection Agency. He is interested in the interface of information, place, the environment, and social and economic justice. A student of group process, he has recently completed training as a certified mediator. His more recent work has been focused on the Earth Charter as a tool for bringing communities together to talk about building more positive futures. A community activist in both his workplace and local area, he aspires to watch the trees grow.

Richard B. Norgaard has been a professor of agricultural and resource economics and of energy and resources at the University of California, Berkeley, since 1970. He felt compelled to study economics after seeing the Glen Canyon of the Colorado River drown beneath the rising waters of Lake Powell in 1963. He was David Brower's (then executive director of the Sierra Club) river guide on seven trips through the Glen, where he also encountered many other well-known environmental activists, artists, and scholars. He earned his bachelor's degree at the University of California at Berkeley, a master's degree at Oregon State University, and a Ph.D. in economics at the University of Chicago. He has lived and undertaken research in the big frontiers of Alaska and the Brazilian Amazon, as well as worked on the biological control of pests in industrialized California agriculture. He is the past president of the International Society for Ecological Economics, generally prefers the company of biologists, and continues to get his raft out on the rivers of the West.

David W. Orr is professor and chair of the environmental studies program at Oberlin College. He is the author of *Earth in Mind, The Nature of Design,* and, most recently, *The Last Refuge.* Having grown up in western Pennsylvania, he has an affinity for rolling hills, good farms, and small towns. With southern roots, he is also a civil war buff. He is a grandfather of two, hence a stakeholder in the farther horizon.

Eric Pallant is a professor of environmental science at Allegheny College. He was chair of Allegheny's Environmental Science Department from 1989 to 1998 and has been director of the Center for Economic and Environmental Development since its creation in 1997. His primary interest is in applying sustainability theory to real-world problems. He has studied, taught, and promoted sustainability in Ecuador, Costa Rica, Meadville, Pennsylvania, and most recently in Israel as a Fulbright scholar. Although he is not always convinced that the ideas of sustainability will prevail over the forces of environmental degradation, he perseveres because doing the right thing, even in the face of defeat, is the only morality he can abide. His courses and CV are on his Web site <http://webpub.allegheny.edu/employee/e/epallant>. He has been making sourdough multigrain breads for nearly two decades because dough feels good and because natural bread smells and tastes better than store-bought bread.

Debra Rowe is the interim dean of applied and engineering technology at Oakland Community College. For the past twenty-two years, she has been professor of environmental systems technology, specializing in renewable energies and energy efficiency. She is also a professor of psychology and has taught solar energy heating, cooking, and water purification to architecture and engineering professors in Oaxaca, Mexico. She holds a Ph.D. in business and two master's degrees from the University of Michigan. Her undergraduate degree is an interdisciplinary major in societal change and change agent skills from Yale. She has just completed a model curriculum about energy for community and technical colleges <http://www.ateec.org/energy>, funded by the U.S. Department of Energy. She has spent the past ten years presenting nationally about interdisciplinary projects in which students work to create a more humane and environmentally sound future <http://www.askeric.org/cgi-bin/printlessons.cgi/Virtual/Lessons/Interdisciplinary/INT0201.html>. She is a senior fellow with the University Leaders for a Sustainable Future.

Paul Rowland is dean of the School of Education and professor of curriculum and instruction at the University of Montana following fourteen years at Northern Arizona University where he served as director of academic assessment, chair of environmental sciences, associate director of the Arizona K-12 Center, and coordinator of environmental education while holding a joint appointment between the College of Education and the College of Arts and Sciences. His research and writing has focused on environmental education, Native American science education, and technology in education. He has taught classes in environmental sciences, curriculum development, higher education, solar technologies, and environmental justice. Sea kayaking (especially on large lakes), cross-country skiing, and hiking the Rockies are among his favored pursuits.

Michael G. Schmidt joined the Medical University of South Carolina (MUSC) in April 1989 after a National Cancer Institute post doctoral training fellowship at the State University of New York at Stony Brook. He is professor and vice chairman of microbiology and immunology, with research interests in bacterial protein export, molecular pathogenesis, environmental microbiology, and microbial community development. Much of his research has been conducted with interdisciplinary teams, so it was a natural step for him to undertake the role of principal investigator for the MUSC component of the Sustainable Universities Initiative. He thinks it only logical to use an interdisciplinary approach to help make integrated systems thinking automatic for faculty and students, producing a more knowledgeable citizenry, capable of making informed choices in order to solve problems quickly and creatively.

Christopher Uhl professor of biology at Pennsylvania State University, had an interest in both medicine and ecology, early in his career, and as his life's work has unfolded, he has been able to join these interests under the rubric of ecological healing. During the 1980s and 1990s, his research focused on the process of ecosystem healing after natural and human-caused disturbances in the Amazon Basin. More recently, he has been directing his attention to the dynamics of biological disturbance and healing in the United States. Using the tools and perspectives of ecology, economics, political science, and ethics, his research and teaching efforts are aimed at promoting sustainable practices and ecological literacy at both local and national levels. Uhl's commitment to ecological healing also permeates his teaching. In his view, the ecological crisis is, to a significant degree, the result of a worldview and concomitant teaching practices that objectify life. With this concern in mind, he has been devising pedagogical approaches that help students see themselves as flesh-and-blood ecological beings embedded in planetary processes.

Polly Walker is the associate director for the Center for a Livable Future at Johns Hopkins Bloomberg School of Public Health. She is a graduate of Radcliffe College and Harvard Medical School. After leaving clinical medicine to raise her family, she worked to improve science and nature education, increase land preservation, protect reservoir watersheds and streams, and increase pollution control. Advocating for these environmental issues out in the field, the disconnect between health and the environment became very apparent to her. These are also health issues: proper stewardship of the environment will prevent many current and future health problems. Her current professional interests are in finding ways to bridge the health-environment gap and to increase sustainable practices. She was a founder and now coordinator of the Johns Hopkins Ad Hoc Committee for the Greening of Johns Hopkins. Two new grandchildren increase the urgency of learning to live sustainably. <http://www.jhsph.edu/environment/>

Index